Effective Leadership

Ten Steps for Technical Professions

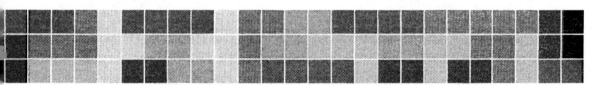

DAVID L. GOETSCH
CEO, Institute for Continual Improvement

NetEffect Series

D1402313

PEARSON
Prentice
Hall

Upper Saddle River, New Jersey
Columbus, Ohio

Library of Congress Cataloging-in-Publication Data

Goetsch, David L.
Effective leadership : ten steps for technical professions / David L. Goetsch.
    p. cm. — (NetEffect series)
Includes bibliographical references and index.
ISBN 0-13-048510-1
1. Leadership. I. Title. II. Series.
HD57.7 .G658 2005
658.4'092—dc22

                                                2003020122

Editor in Chief: Stephen Helba
Executive Editor: Debbie Yarnell
Associate Editor: Kimberly Yehle
Production Editor: Louise N. Sette
Production Supervision: Gay Pauley, Holcomb Hathaway
Design Coordinator: Diane Ernsberger
Cover Designer: Ali Mohrman
Production Manager: Brian Fox
Marketing Manager: Jimmy Stephens

Pearson Education Ltd.   Pearson Education Australia Pty. Limited
Pearson Education Singapore Pte. Ltd.   Pearson Education North Asia Ltd.
Pearson Education Canada, Ltd.   Pearson Educación de Mexico, S.A. de C.V.
Pearson Education—Japan   Pearson Education Malaysia Pte. Ltd.

ISBN 0-13-048510-1

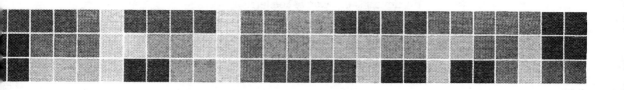

Contents

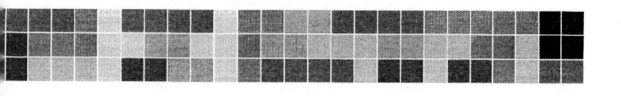

Introduction:
Effective Leadership–
An Overview

"I know even better than I used to that a lot of U.S. businesses are sinking in a sea of bureaucracy. What many need is a new skipper."

—Carl Icahn
Trans World Airlines

OBJECTIVES

- Define effective leadership.
- Explain whether leadership skills are inherited or learned.
- Explain the relationship between motivation and inspiration.
- Explain the different leadership styles.
- Describe the best leadership style for a competitive setting.
- Explain how to build and maintain a following.
- Describe the differences between leadership and management.
- Explain the role of ethics in leadership.

Leadership is an intangible concept that produces tangible results. It is referred to sometimes as an art and at other times as a science. In reality, effective leadership is both an art and a science.

The impact of good leadership can be readily seen in any organization where it exists. Well-led organizations, whether they are large or small, share several easily identifiable characteristics:

- High levels of productivity
- Positive, can-do attitudes
- Commitment to accomplishing organizational goals
- Effective, efficient use of resources
- High levels of quality
- Mutually supportive teamwork approach to getting work done

Where effective leadership exists, work is accomplished primarily by teams. These teams are built deliberately, nurtured carefully, and improved continually through effective leadership. This introduction provides an overview of the concept of leadership and how it applies in a globally competitive environment.

EFFECTIVE LEADERSHIP DEFINED

Leadership can be defined in many different ways, partly because it has been examined from the perspective of so many different fields of endeavor. Leadership has been defined as it applies to the military, athletics, education, business, industry, and many other fields. For the purposes of this book, effective leadership is defined as it relates specifically to technology professionals:

Effective leadership is the ability to inspire people to make a total, willing, and voluntary commitment to accomplishing or exceeding organizational goals.

This definition contains a key concept that makes it particularly applicable in globally competitive environments: the concept of inspiring people. Inspiring people is a higher order of human interaction than motivating, which is a concept more frequently used in defining leadership. Motivated employees commit to the organization's goals. Inspired employees make those goals their own. When employees are inspired, the total, willing, and voluntary commitment described in the definition follows naturally.

Characteristics of an Effective Leader

Effective leaders come in all shapes, sizes, genders, ages, races, political persuasions, and national origins. They do not look alike, talk alike, or even work alike. However, effective leaders do share several common characteristics. These are the characteristics necessary to inspire people to make a total, willing, and voluntary commitment. Regardless of their backgrounds, effective leaders exhibit the characteristics shown in Figure I.1.

Characteristics of Effective Leaders

 ✓ Persuasiveness
 ✓ Positive influence
 ✓ Good communication skills
 ✓ Positive role model
 ✓ Balanced commitment

FIGURE I.1   Characteristics of effective leaders.

Effective leaders are committed to both the job to be done and the people who must do it, and they are able to strike the appropriate balance between the two. Effective leaders project a positive example at all times. They are good role models. Leaders who project a "Do as I say, not as I do" attitude will not be effective leaders. To inspire employees, leaders must be willing to do what they expect of workers, do it better, do it right, and do so consistently. If, for example, dependability is important, leaders must set a consistent example of dependability. If punctuality is important, a leader must set a consistent example of punctuality. To be an effective leader, a technical professional must set a consistent example of all characteristics that are important on the job.

Effective leaders are good communicators. They are willing, patient, and skilled listeners. They are also able to communicate their ideas clearly, succinctly, and in a nonthreatening manner. They use their communication skills to establish and nurture rapport with employees. Effective leaders have influence with employees and use it in a positive manner. Influence is the art of using power to move people toward a certain end or point of view. The power of leaders derives from the authority that goes with their jobs and the credibility they establish by being effective leaders. Power is useless unless it is converted to influence. Power that is properly, appropriately, and effectively applied becomes positive influence.

Finally, effective leaders are persuasive. Leaders who expect people to simply do what they are ordered to do will have limited success. Those who are able to use their communication skills and influence to persuade people to their point of view and to help people make a total, willing, and voluntary commitment to that point of view can have unlimited success.

Myths about Leadership

Over the years a number of myths have grown up about the subject of leadership. Technical professionals should be aware of these myths and be able to dispel them. In Leaders: The Strategies for Taking Charge, W. Bennis and B. Nanus describe the most common myths about leadership as follows.[1] (The author dispels each myth as it is presented.)

Myth 1: Leadership is a rare skill.   Although it is true that few great leaders of world renown exist, many good, effective leaders do. Renowned leaders such as Winston Churchill were simply good leaders given the opportunity to participate in monumental events (World War II in Churchill's case). Another example is General Norman Schwarzkopf. He had always been an effective military leader. That's how he became a general. But it took a monumental event—the Gulf War—coupled with his leadership ability to make General Schwarzkopf a world-renowned leader. His leadership skills did not appear suddenly; he had them all along. Circumstances allowed them to be displayed on the world stage. Most effective leaders spend their careers in virtual anonymity, but they exist in surprisingly large numbers, and there may be little or not correlation between their ability to lead and their relative positions in an organization. The best leader in a company may be the lowest-paid wage earner, and the worst may be the CEO. In addition, a person may be a leader in one setting and not in another. For example, a person who shows no leadership ability at work may be an effective leader in his church. One of the keys to success is to create an environment that brings out the leadership skills of all employees at all levels and focuses them on continually improving competitiveness.

Myth 2: Leaders are born, not made.   This myth is addressed later in this chapter. Suffice it to say here that leadership attitudes and behaviors can be learned, even by those who do not appear to have inborn leadership potential.

Myth 3: Leaders are charismatic.   Some leaders have charisma and some do not. Some of history's most renowned leaders had little or no charisma. Correspondingly, some of history's greatest misleaders were highly charismatic. Generals Dwight Eisenhower and Omar Bradley are examples of great but uncharismatic leaders. Adolf Hitler and Benito Mussolini are examples of great misleaders who relied almost exclusively on charisma to build a following.

Myth 4: Leadership exists only at the top.   Organizations could not be competitive if this were true. Competitiveness relies on the building of teams at all levels in an organization and teaching employees in these teams to be leaders. In reality, the opposite of this myth is often true. Top leaders may be the least-capable leaders in a company. Leadership is about producing results and generating continual improvement, not one's relative position within the organization.

Myth 5: Leaders control, direct, prod, and manipulate. If practice is an indicator, this myth is the most widely believed. The "I'm the boss; do what I say" syndrome is rampant in business and industry. It seems to be the automatic fallback position or default approach for managers who don't know better. Leadership in a competitive environment is about involving and empowering, not prodding and manipulating.

Myth 6: Leaders do not need to be learners. Lifelong learning is a must for leaders. One cannot be an effective leader without being a good learner. Leaders do not learn simply for the sake of learning (although to do so is a worthwhile undertaking). Rather, leaders continually learn in what Bennis and Nanus call an "organizational context."[2] This means they approach learning from the perspective of what matters most to their organizations. A leader who is responsible for the technical writing department in a software production company might undertake to learn more about the classics of European literature. Although this would certainly make her a better-educated person, studying European literature is not learning in the organizational context of the leader of a technical writing department. Examples of learning in an organizational context for such a leader include learning new software programs, enrolling in pertinent technical writing classes, and staying abreast of the latest trends in software development.

LEADERSHIP SKILLS: INHERITED OR LEARNED?

One of the oldest debates about leadership revolves around the question of whether leaders are born or made. Can leadership skills be learned, or must they be inherited? This debate has never been settled and probably never will be. There are proponents for both sides of the issue, and this polarity is not likely to change, because, as is often the case in such controversies, both sides are partially right.

Leaders are like athletes: some are born with natural ability, whereas others develop their ability through determination and hard work. Inborn ability, or the lack of it, represents only the starting point. Success from that point forward depends on the individual's willingness and determination to develop and improve. Some athletes born without tremendous natural ability do, through hard work, determination, and continuous improvement, perform beyond their apparent potential.

This phenomenon also applies to leadership. Some people have more natural leadership ability than others do. However, regardless of their individual starting points, people can become good leaders through education, training, practice, determination, and effort.

LEADERSHIP, MOTIVATION, AND INSPIRATION

One of the characteristics shared by effective leaders is the ability to inspire and motivate others to make a commitment. The key to motivating employees lies in the ability to relate their personal goals to the company's goals. The key to inspiring employees lies in the ability to relate what they believe in to the company's goals. Implicit in both cases is the leader's need to know and understand employees, including both their individual needs and their personal beliefs.

Understanding Individual Needs

Perhaps the best model for explaining individual human needs is that developed by psychologist Abraham H. Maslow. Maslow's Hierarchy of Needs (Figure I.2) arrays the basic human needs on five successive levels. The lowest level in the hierarchy encompasses basic survival needs. All people need air to breathe, food to eat, water to drink, clothing to wear, and shelter from the elements. The second level encompasses safety and security needs. All people need to feel safe from harm and secure in their world. To this end, people enact laws, pay taxes to employ police and military personnel, buy insurance, try to save and invest money, and install security systems in their homes.

The third level encompasses social needs. People are social animals by nature. This fact manifests itself through families, friendships, social organizations, civic groups, special clubs, and even employment-based groups such as company softball and basketball teams. The fourth level of the hierarchy encompasses esteem needs. Self-esteem is a key ingredient in the per-

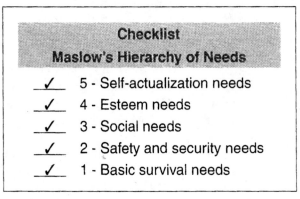

FIGURE 1.2   Maslow's Hierarchy of Needs.

sonal happiness of individuals. All people need to feel self-worth, dignity, and respect. People need to feel that they matter. This fact manifests itself in a variety of ways. It can be seen in the clothes people wear, the cars people drive, and the behavior people exhibit in public.

The highest level of Maslow's hierarchy encompasses self-actualization needs. Complete self-fulfillment is a need that is rarely satisfied in people. The need for self-actualization manifests itself in a variety of ways. Some people seek to achieve it through their work; others through hobbies, relationships, or leisure activities.

Leaders need to understand how to apply Maslow's model if they hope to use it to motivate and inspire workers. Principles required for applying this model are as follows:

1. Needs must be satisfied in order from the bottom up (most basic to highest).
2. People focus most intently on their lowest unmet need. For example, employees who have not met their basic security needs will not be motivated by factors relating to their social needs.
3. After a need has been satisfied, it no longer works as a motivating factor. For example, employees who have satisfied their need for financial security will not be motivated by a pay raise.

Understanding Individual Beliefs

Every person has basic beliefs that, when taken together, form that individual's value system. If leaders know their fellow employees well enough to understand those basic beliefs, they can use this knowledge to inspire them on the job. Developing this level of understanding of employees comes from observing, listening, asking, and taking the time to establish trust.

Leaders who develop this level of understanding of employees can use it to inspire them to higher levels of performance. This is done by showing employees how the organization's goals relate to their beliefs. For example, if pride of workmanship is part of an employee's value system, a leader can inspire that person to help achieve the organization's goals by appealing to that value.

Inspiration, as a level of leadership, is on a higher plane than motivation. Technical professionals who become good enough leaders to inspire their employees will achieve the best results.

LEADERSHIP STYLES

Leadership styles have to do with how people interact with those they seek to lead. Leadership styles go by many different names. However, most styles fall into one of five categories: autocratic, democratic, participative, goal-oriented, or situational.

Autocratic Leadership

Autocratic leadership is also called "directive" or "dictatorial" leadership. People who take this approach make decisions without consulting the employees who will have to implement them or who will be affected by them. They tell others what to do and expect them to comply obediently. Critics of this approach say that although it can work in the short run or in isolated instances, in the long run it is not effective.

Democratic Leadership

Democratic leadership is also called "consultative" or "consensus" leadership. People who take this approach involve the employees who will have to implement decisions in making them. The leader actually makes the final decision, but only after receiving the input and recommendations of team members. Critics of this approach say the most popular decision is not always the best decision and that democratic leadership, by its nature, can result in the making of popular decisions as opposed to right decisions. This style can also lead to compromises that ultimately fail to produce the desired result.

Participative Leadership

Participative leadership is also known as "open" or "nondirective" leadership. People who take this approach exert only minimal control over the decision-making process. They provide information about the problem and allow team members to develop strategies and solutions. The underlying assumption of this style is that people will more readily accept responsibility for solutions, goals, and strategies they are empowered to help develop. Critics of this approach say it is time-consuming and works only if all people involved are committed to the best interests of the organization.

Goal-Oriented Leadership

Goal-oriented leadership is also called "results-based" or "objective-based" leadership. People who take this approach ask team members to focus solely on the goals at hand. Only strategies that make a definite and measurable contribution to accomplishing organizational goals are discussed. The influence of personalities and other factors unrelated to the specific goals of the organization are minimized. Critics of this approach say it can break down when team members focus so intently on specific goals that they overlook opportunities or potential problems that fall outside of their narrow focus. Results-oriented leadership can be too narrowly focused and centered on the wrong concerns.

Situational Leadership

Situational leadership is also called "fluid" or "contingency" leadership. People who take this approach select the style that seems to be appropriate given the circumstances at the time. In identifying these circumstances, leaders consider the following factors:

- Relationship of the leader and team members
- How precisely actions taken must comply with specific guidelines
- Amount of authority the leader has with team members

Depending on what is learned when these factors are considered, the leader decides whether to take the autocratic, democratic, participative, or goal-oriented approach. Under different circumstances, the same person would apply a different leadership style. Critics reject situational leadership as an approach based on short-term concerns instead of on solutions to long-term problems.

LEADERSHIP STYLE IN A GLOBALLY COMPETITIVE SETTING

The appropriate leadership style in a global setting might be called participative leadership taken to a higher level. Whereas participative leadership in the traditional sense involves soliciting employee input, in a global setting the employees giving the input must be empowered. In addition, employee input must be taken seriously. The key differences between the traditional participative leadership and participative leadership as advocated by the author are that, with the latter, employees providing the input are empowered and their input is given serious consideration.

Collecting employee input is not new. However, collecting input, logging it in, tracking it, acting on it in an appropriate manner, working with employees to improve weak suggestions rather than simply rejecting them, and rewarding employees for improvements that result from their input extend beyond the traditional approach to participative leadership.

BUILDING AND MAINTAINING A FOLLOWING

Technical professionals can be good leaders only if the people they hope to lead will follow them willingly and steadfastly. Followership must be built, and having been built, maintained. Following is a discussion of how technical professionals can build and maintain followership among the people they hope to lead.

Popularity and the Leader

Leadership and popularity are not the same thing. However, many people confuse popularity with both leadership and followership. An important point to understand in leading people is the difference between popularity and respect. Long-term followership grows out of respect, not popularity. Good leaders may be popular, but they must be respected. Not all good leaders are popular, but they are all respected.

Technical professionals occasionally have to make unpopular decisions. This is a fact of life for leaders, and it is why leadership positions are sometimes described as lonely ones. Making an unpopular decision does not necessarily cause a leader to lose followership, provided the leader is seen as having solicited a broad base of input and given serious, objective, and impartial consideration to that input. Correspondingly, leaders who make inappropriate decisions that are popular in the short run may actually lose followership in the long run. If the long-term consequences of a decision turn out to be detrimental to the team, team members will hold the leader responsible, particularly if the decision was made without first collecting and considering employee input.

Leadership Characteristics That Build and Maintain Followership

Leaders build and maintain followership by earning the respect of those they lead. The following characteristics of leaders build respect:

■ Sense of purpose and vision. Successful leaders have a strong sense of purpose. They know who they are, where they fit in the overall organization, and the contributions their areas of responsibility make to the organization's success.

■ Self-discipline. Successful leaders develop discipline and use it to set an example. Through self-discipline, leaders avoid negative self-indulgence, inappropriate displays of emotion such as anger, and counterproductive responses to the everyday pressures of the job. Through self-discipline, leaders set an example of handling problems and pressures with equilibrium and a positive attitude.

■ Honesty and integrity. Successful leaders are trusted by their followers. This is because they are open, honest, and forthright with other members of the organization and with themselves. They can be depended on to make difficult decisions in unpleasant situations with steadfastness, consistency, and integrity.

■ Credibility. Successful leaders have credibility. Credibility is established by being knowledgeable, consistent, fair, and impartial in all human interactions; by setting a positive example; by adhering to the same standards of performance and behavior expected of others; and by being a good steward with the organization's resources.

■ Stamina. Successful leaders must have stamina. Frequently they need to be the first to arrive and the last to leave. Their hours are likely to be longer and the pressures they face more intense than those of others. Energy, endurance, and good health are important to those who lead.

■ Commitment. Successful leaders are committed to the goals of the organization, the people they work with, and their own ongoing personal and professional development. They are willing to do everything within the limits of the law, professional ethics, and company policy to help their team succeed.

■ Steadfastness. Successful leaders are steadfast, resolute, and willing to persevere. People do not follow a person they perceive to be wishy-washy and noncommittal. Nor do they follow a person whose resolve they question. Successful leaders must have the steadfastness to stay the course even when it becomes difficult.

LEADERSHIP AND ETHICS

In his book Managing for the Future, Peter Drucker discusses the role of ethics in leadership.[3] Setting a high standard of ethical behavior is an essential task of leaders in a global setting. Drucker summarizes his view regarding ethical leadership in the modern workplace as follows:

> What executives do, what they believe and value, what they reward and whom, are watched, seen, and minutely interpreted throughout the whole organization. And nothing is noticed more quickly—and considered more significant—than a discrepancy between what executives preach and what they expect their associates to practice. The Japanese recognize that there are really only two demands in leadership. One is to accept that rank does not confer privileges; it entails responsibilities. The other is to acknowledge that leaders in an organization need to impose on themselves that congruence between deeds and words, between behavior and professed beliefs and values, that we call personal integrity.[4]

Drucker speaks to ethical behavior of corporate executives who hope to lead their companies effectively. The concept also applies to technical professionals at all levels as well as any other person who hopes to lead others in a global setting.

EFFECTIVE LEADERSHIP: TEN STEPS FOR TECHNICAL PROFESSIONALS

This book uses a "great leaders" approach to help technical professionals develop the knowledge, skills, insights, and attitudes they need to be effective leaders. This approach works as follows: (1) 10 fundamental characteristics shared by selected individuals who are widely viewed as having been great leaders are presented in 10 respective chapters; (2) within each chapter one of the fundamental characteristics and what must be known about it are set forth and then illustrated through the life and work of several great leaders; and (3) several leadership simulations are provided at the end of each chapter to generate thought, discussion, and debate concerning how the leadership characteristic under discussion applies in today's globally competitive workplace.

The 10 steps to becoming an effective leader set forth in this book are as follows:

1. Develop a vision and commit to achieving it.
2. Project unquestionable integrity and selflessness.
3. Establish credibility and good stewardship.
4. Develop a can-do attitude and seek responsibility.
5. Develop self-discipline, time management, and execution skills.
6. Be a creative problem solver and decision maker.
7. Be a positive change agent.
8. Be an effective team builder.
9. Empower followers to lead themselves.
10. Be an effective conflict manager and consensus builder.

The great leaders selected to illustrate the various leadership characteristics set forth in this book come primarily from the worlds of business, sports, politics, government, and the military. Leadership characteristics are universally applicable. Consequently, sometimes the best example of the application of a given characteristic might come from a seemingly unrelated field (e.g., a military example that applies well in a business or technology setting). The reader is encouraged to think beyond the narrow scope of the field in question to the larger principle being presented and how that principle might be applied in his specific setting.

Rationale for the Great Leaders Approach

Why use leaders from such fields as politics, the military, and sports to teach leadership to technical professionals from the world of business? Why not just use examples taken from the world of business, such as Bill Gates of

Microsoft or other groundbreaking business entrepreneurs? These are good questions, and readers should understand the answers before proceeding further.

Using leaders from outside the world of commerce certainly might seem to some a questionable approach, at least on the surface. After all, leaders from business and commerce work in a world of profit and loss and bottom lines. In contrast, politics often seems murky, inefficient, and disorderly. The military is certainly more disciplined than politics, but military leaders are more concerned with winning battles than with winning customer loyalty. And leaders from the world of sports are concerned with winning games and championships.

So the question remains: What is the rationale for using the great leaders approach in a book for technical professionals from the world of commerce? The answer to this question is very simple. Although there certainly are many surface-level differences between leadership in the fields of business and commerce and leadership in the fields of politics, the military, and sports, a look beneath the surface reveals some interesting and even striking parallels. These parallels were created by, or at least magnified by, the concept "market globalization."

To lead in the fields of politics, the military, and sports, one must confidently and firmly project a strong sense of purpose and direction. This is also true of those who would lead an organization competing in the global marketplace. To win campaigns, battles, and championships, leaders from the fields of politics, the military, and sports must become adept at developing and implementing strategies. This is also true of technical professionals who would lead their organizations to victory in the global marketplace. In fact, leaders in some Japanese companies are taught that "business is war."

Finally, leadership in any field is about strength of character more than anything else. Nowhere are the strengths and weaknesses of leaders tested so frequently or so publicly as in the fields of politics, the military, and sports. The qualities of leaders in these fields are played out on a grand stage in the daily, unforgiving glare of close media scrutiny. Consequently, the leaders' successes and their failures have been studied carefully, analyzed thoroughly, and are documented fully. As a result, the lives and experiences of great leaders from the fields of politics, the military, and sports hold invaluable lessons for technical professionals who hope to lead their organizations to victory in an increasingly competitive global marketplace.[5]

Endnotes

1. W. Bennis and B. Nanus, *Leaders: The Strategies for Taking Charge* (New York: Harper & Row, 1985), 222–226.

2. W. Bennis and B. Nanus, 189.

3. Peter F. Drucker, *Managing for the Future* (New York: Harper & Row, 1985), 113–117.

4. Peter F. Drucker, 116–117.

5. Steven F. Hayward, *Churchill on Leadership* (Rocklin, CA: Forum, 1977), xvi–xix.

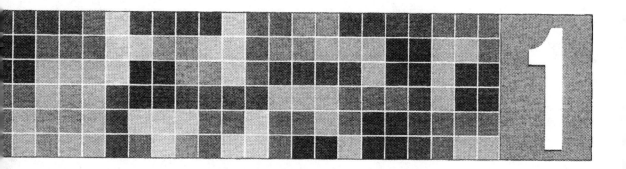

Develop a Vision and Commit To Achieving It

"The greater thing in this world is not so much where we stand but in what direction we are going."[1]

—Oliver Wendell Holmes

OBJECTIVES

- Define the concept of vision.
- Explain the importance of having a clear vision.
- Demonstrate how to develop your vision.
- Explain how to win others to your vision.
- Explain the factors that can hinder development of your vision.
- Summarize the vision lessons of selected effective leaders.

Implicit in the concept of leadership is direction. To lead, you must have a sense of direction, or, said another way, you must know where you are going. People are more apt to follow someone who knows where she is going and is committed to getting there—someone who has a sense of purpose—than someone who lacks direction and commitment.

People who know where they are going are said to have a "vision." The concept of vision encompasses direction, purpose, and commitment. Having a vision and committing yourself to achieving it are fundamental prerequisites to becoming an effective leader. Throughout history, in all fields of endeavor, effective leaders have shared at least three common characteristics: they know where they are going, they are committed to getting there, and they are able to inspire others to go with them.

VISION DEFINED

A vision is a picture of the leader's intended destination for the organization she hopes to lead. That destination can be described as what the leader wants her organization to be or where she wants it to go. Regardless of the size of the organization, be it a company, division, department, unit, or team, a vision must convey the leader's intended destination clearly, simply, and understandably. A good vision has the following characteristics:

1. Easily understood by all stakeholders
2. Briefly stated, yet clear and comprehensive in meaning
3. Challenging, yet attainable
4. Lofty, yet tangible
5. Capable of inspiring commitment among all stakeholders
6. Capable of setting the tone for stakeholders
7. Able to establish direction without getting into numbers (numbers are reserved for goals and objectives)

IMPORTANCE OF A VISION

A well-developed vision gives more than just *direction* to employees; it also gives *meaning* to the work they do every day (see Figure 1.1). By showing them a brighter future, a leader with a vision can help employees as they struggle with the day-to-day realities of working in a globally competitive environment. This issue of work having meaning is important because it is easy for employees to get so bogged down in the daily grind and the stress of getting the job done that they lose their *focus* and begin to feel that their work, which is a large part of their lives, has no meaning. Employees who think their work is meaningless are not likely to perform at their best day

> **Benefits of a**
> **Well-Developed Vision**
>
> - Provides *direction.*
> - Keeps employees *focused* on a brighter future.
> - *Inspires* employees to perform at their best when work becomes frustrating.
> - Gives *meaning* to the work lives of employees.

FIGURE 1.1   A well-crafted vision can serve as a rallying point for employees.

in and day out, nor are they likely to invest the time, attention, and effort necessary to improve their performance on a continual basis.

Another reason the vision is so important is that the road to success in any business is typically littered with so many potholes, detours, and other obstacles that employees can become frustrated and just give up. A well-developed vision can serve as a beacon in the distance that *inspires* employees to persevere in spite of the obstacles. It can serve the same purpose that a clearly visible finish line in the distance serves for marathon runners who are struggling with exhaustion, sore feet, and muscle cramps.

In *The Leadership Challenge*, James Kouzes and Barry Posner summarize the importance of having a vision as a leadership strategy: "One of *the* most important practices of leadership is giving life and work a sense of meaning and purpose by offering an exciting vision."[2]

HOW TO DEVELOP A VISION

The process of developing a vision (see Figure 1.2) is much the same for technical professionals regardless of whether the organization they hope to lead is a company, division, department, unit, or team. The major difference between a vision for a company and any smaller unit of organization is the context within which the vision is developed. Following is a step-by-step explanation of how to develop a vision.

Leadership Tip

"The horizon is out there somewhere, and you just keep chasing it, looking for it, working for it."[3]

—Robert Dole
U.S. Senator

Developing a Vision

1. Study *visionary leaders* and consider the *possible.*
2. Establish your *context.*
3. Examine the *present.*
4. Study the *future.*
5. Identify the *ideal condition* within your context.
6. Ask the *legacy* question.
7. *Write* your vision statement.
8. *Test* the vision statement.

FIGURE 1.2   Developing a vision is a step-by-step process.

Study Visionary Leaders and Consider the Possible

The natural tendency of people who are first learning to develop a vision is to think too small or too conventionally. They tend to erect self-imposed fences that limit them and the organization they hope to lead. By studying visionary leaders, you can learn to consider the *possible.* This is sometimes referred to as thinking outside of the box. Great leaders become great leaders in part by establishing great visions. People who think small generally stay small. The visions of several great leaders are presented later in this chapter.

Establish Your Context

A vision is, by its very nature, a qualitative concept. It should be stated in such a way that if it is achieved, the organization will *win* in the marketplace chosen as its arena. Qualitatively you should envision an organization that is the best at what it does in the world, country, region, or some other identifiable market segment. If your vision is for a company, the context for the vision may be global, national, regional, or local, but whatever it is, the context must be defined. The choice of how high the leader chooses to aim—that is, how broadly she chooses to define the competition—can make a significant difference in the strategies adopted as her company pursues its vision. The strategies needed to be the leading company in a global market will be radically different and much more challenging than those needed to win in the local marketplace.

The leader must strike an appropriate balance between "challenging" and "attainable" when deciding how high to aim with a vision. Aim too high and the vision becomes unattainable and, therefore, unrealistic in the minds of employees. Remember, a well-crafted vision is challenging, yet attainable. What is challenging but still attainable depends on the organization's current status vis-à-vis the competition. For example, two engineers who just invested their life savings in establishing a new engineering firm that has a total of five employees would not want to choose a global vision at the outset. Instead, they might begin by focusing on becoming the premier engineering firm in their local community or on providing a limited number of services, or both. Then, as they make progress in pursuing this relatively narrow vision, they might begin to broaden it—first to the regional level, then to the national level, and ultimately to the global level.

It is just as inappropriate to aim too low when developing a vision. Organizations that make this mistake risk "underwhelming" their employees with lack of a good challenge and boring them with insufficient loftiness.

If the organization is smaller than a company—a division, department, unit, or team—the context is established by the next larger segment of the company into which the organization fits. For example, if the organization is a division, its leader would develop a vision that fits into the context of the company (e.g., to be the most productive division in the company). If the organization is a work team, its leader would develop a vision that fits into the context of the department that houses the team (e.g., to be the highest-rated team in the department in quality, productivity, and customer satisfaction).

Examine the Present

Although a vision looks to the future, it is a good idea to conduct a thorough examination of the present when developing a vision. This is how the leader ensures that the final vision is not just lofty, but tangible, and not just challenging, but attainable. She should ask questions such as: Where are we right now in terms of the competition in our market? What organization is currently the best? How far ahead of us are they? How are they different from us? How broad can our market or performance context be without appearing to be unrealistic to employees?

Study the Future

At the beginning of World War II, the allied countries of France and Poland were quickly overrun by Adolf Hitler's futuristic approach to war that came to be known as the blitzkrieg, or lightning war. While the Poles attempted to defend themselves with cavalry and the French waited in the static positions they employed during World War I (the Maginot Line), Hitler's troops aggressively applied new forward-looking tactics involving the coordinated use of mobilized infantry, tanks, and aircraft in a devastating assault that quickly went over, through, or around even the best of the outmoded Polish and French defenses. Historically, military leaders have often been guilty of looking backward and trying to fight the next war with outdated tactics

---

**Resources for
Studying the Future**

- *The 500-Year Delta: What Happens after What Comes Next*
  by Watts Wacker and Jim Taylor, with Howard Means
  (New York: HarperBusiness, 1997)
- *The Art of the Long View: Planning for the Future in an
  Uncertain World* by Peter Schartz (New York:
  Doubleday/Currency, 1991)
- *Thinking in the Future Tense: Leadership Skills for a New Age*
  by Jennifer James (New York: Simon & Schuster, 1996)
- *Where on Earth Are WE Going?* by Maurice Strong
  (New York: Texere, 2001)

FIGURE 1.3   To consider the possible requires a thorough study of the future.

from the last. Technical professionals who hope to lead an organization in a global setting cannot afford to make the same mistake.

When developing your vision, take some time to study the future, and make a point of looking forward, not backward. Figure 1.3 contains several resources that can help you expand your mental horizons beyond just the here and now and avoid the mistake of limiting your thinking to the everyday realities of the present.

Identify the Ideal Condition Within Your Context

You are attempting to lead an organization of a given size in the battle of the marketplace. The context you established earlier defines the scope of the marketplace in which you hope to compete. Within that context, if you could simply wave a magic wand and make it happen, what condition would you create? What is the ideal picture of your organization within its established marketplace?

Would you like to be the market leader? Would you like to be in the top 10 percent in your market? Would you like to be the leader in a specific market niche within the broader market? Would you like your branch plant to be the highest-rated plant in the overall company? Would you like your department to be the most productive in your division or within the entire company? Would you like your team to have the highest quality rating of any team in your company or the highest customer satisfaction rating, or both?

When deciding on the ideal condition, don't be afraid to aim high. Even if the ideal condition is unrealistic at the moment, it may not always be. Visions can be set at one level in line with current circumstances and increased to higher levels as improved performance allows.

Ask the Legacy Question

Your legacy is how you will be remembered when you are gone. When developing a vision, it can be helpful to ask yourself, "How do I want the organization I led to be remembered?" This is similar to identifying the ideal condition discussed in the previous section. You might want people to remember that under your leadership the organization became the most productive, most efficient, most competitive, or most anything else organization it has ever been. Your desired legacy can become your vision or a major part of it.

Write Your Vision Statement

Developing a vision is just the first step in a process that also involves communicating the vision to stakeholders and winning their support for it. To communicate your vision, it is necessary to convert it into a written statement. Earlier in this chapter, the seven characteristics of a good vision were presented. These characteristics apply not just to the vision but also to the vision statement. When writing your vision statement, keep the following characteristics in mind:

■ *Easily understood by all stakeholders.* The stakeholders to whom your vision statement should be directed include any person or group that will play any kind of role, direct or indirect, in moving your organization toward accomplishment of its vision. Before beginning to write your vision statement, make a list of all stakeholders. This list is the audience for what you will write.

■ *Briefly stated, but clear and comprehensive in meaning.* A well-written vision statement may not fit on a bumper sticker, but it should fit on the back of a business card. Anything longer than a brief paragraph is too long. Every stakeholder should be able to easily memorize the vision statement. It should be long enough to clearly and comprehensively convey its meaning, and no longer. This can take a lot of practice. It will probably be the most challenging characteristic to incorporate in your vision statement.

■ *Challenging, yet attainable.* If you aim too low when developing a vision, employees may be insufficiently challenged and not take the vision seriously. If you aim too high, employees may simply become overwhelmed and give up. Striking an appropriate balance is a matter of developing a statement that conveys the message that the vision is currently beyond the organization's grasp but not permanently out of reach. A vision statement that strikes an appropriate balance is one about which employees can say, "It will be difficult, but we can do it."

■ *Lofty, yet tangible.* People are generally idealistic by nature. They respond to challenges that put them on the high road to accomplishing something worthwhile. Abraham Lincoln knew this when he wrote the Gettysburg Address. When he delivered this brief but powerful speech, the American Civil War had dragged on much longer than either side had originally thought it would, and thousands of families on both sides had lost loved ones. Many in the North had tired of the war and were ready to just let the South go its separate way—and then came Gettysburg. This horrible slaughter near the small Pennsylvania town that gave the battle its name was a victory for the North, but a costly one and not the decisive victory it could have been had Lincoln gotten better leadership from General Meade, the latest in a long list of underperforming Union generals. Lincoln knew that victory could be had, but only if he could rally the flagging support of Northern citizens to his cause. He needed to give them a vision that was lofty enough to convince them to continue the fight, yet tangible enough that the common citizen could grasp its feasibility. Lincoln gave the North—and ultimately the world—the Gettysburg address, one of the most effectively written visions ever committed to paper:

> Four score and seven years ago our fathers brought forth on this continent a new nation, conceived in Liberty, and dedicated to the proposition that all men are created equal. We are now engaged in a great civil war, testing whether that nation, or any nation so conceived and so dedicated, can long endure. We are met on a great battlefield of that war. . . . It is for us the living, rather, to be dedicated here to the unfinished work which they who fought here have thus far so nobly advanced. It is rather for us to be here dedicated to the great task remaining before us—that from these honored dead we take increased devotion to that cause for which they gave the last full measure of devotion—that we here highly resolve that these dead shall not have died in vain—that this nation, under God, shall have a new birth of freedom—and that government of the people, by the people, and for the people, shall not perish from the earth.[4]

Lincoln's words were lofty indeed. He told his stakeholders that they were fighting for nothing less than liberty, freedom, and government of, by, and for the people. But he also showed them that what remained to be done was tangible and could be achieved if they would simply persevere in the spirit of those who had already given the "last full measure of devotion."

■ *Capable of inspiring commitment among all stakeholders.* To win the commitment of stakeholders, a vision must be inspiring—it must be something they will feel good about pursuing. This means that the vision statement must convey a vision that is worth pursuing and that stakeholders can see that it will make a positive difference in their lives (e.g., ensure their job security, provide continued career advancement opportunities, benefit society).

■ *Capable of setting the tone for stakeholders.* Stakeholders who read your vision statement should come away realizing that you are serious about this. The vision statement should convey that the leaders of the organization are determined to succeed and that they mean business.

■ *Able to establish direction without getting into numbers.* The vision statement must tell stakeholders where the organization is going and what it hopes to become, but it must do this without getting into specific numbers. Numbers such as a 50 percent improvement in customer satisfaction or a 20 percent reduction in downtime are reserved for the organization's goals. They have no place in the vision.

Having considered the various characteristics of a good vision statement, you now begin the task of drafting the statement. Following are some examples for organizations of various types and sizes.

■ ABC Company will be the premier provider of civil, structural, electrical, and mechanical engineering services in northwest Florida.
■ The structural engineering department of ABC Company will be the leading department in the company in terms of productivity, efficiency, quality, and customer service.
■ The commercial buildings team of the structural engineering department of ABC Company will consistently lead all other teams in the department in all applicable performance standards.

These examples show the respective visions for a company, a department in that company, and a team within that department. The context for the company's vision is the geographic region of northwest Florida. The context for the department's vision is the company, and the context for the team's vision is the department which houses it.

Test the Vision Statement

Having written a draft of your vision statement, you should now test the draft. Does the statement have all of the characteristics of a good vision statement? To get a feel for how this step is accomplished, consider the following successive drafts of a vision statement for a technology company.

1. XYZ Manufacturing and Engineering Company will be the best in its field.
2. XYZ Manufacturing and Engineering Company will be the best in the United States at providing services in its field.

3. XYZ Manufacturing and Engineering Company will be the best in the United States at providing electronics engineering and manufacturing services.
4. XYZ Manufacturing and Engineering Company will be the leading provider of electronics engineering and manufacturing services for low-voltage power supplies in the southeastern United States.

These various drafts were developed by the CEO of XYZ Manufacturing and Engineering Company and his executive-level managers. After developing the first draft, the team tested it against the characteristics of a good vision and determined that it needed work. Stakeholders could understand from the draft that the company wants to be the best in its field, but in what context? Best in the world? Best in the United States? The team of executives went back to the drawing board and produced the second draft. This draft was better. It clearly established that the company wants to be the best in the United States in its field.

However, after reflecting on the second draft, the CEO suggested that the team be even more specific and define the company's field for future readers of the vision. The third draft was even more understandable, but one of the team members made the point that there are thousands of different products that require engineering and manufacturing services in the broad field of electronics. She proposed that their company's niche in this field be specified in the vision statement. Her input led to the fourth and final draft.

This final draft has all of the characteristics set forth earlier in this chapter. It can be easily understood by all stakeholders. The statement is brief—just one sentence—yet it is clear and comprehensive. Seeking to be the "leading provider" in any field is certainly challenging, but by putting this in the context of the United States rather than the world, the CEO and his executive managers made the vision attainable considering the company's current circumstances.

Committing to being the best in your field is certainly a lofty undertaking, but by limiting the vision by both product (low-voltage power supplies) and geography (the United States), the development team ensured that the vision was also tangible. Surely people working at XYZ Manufacturing and Engineering Company will be inspired by this final vision. Who would not want to be part of a winner? Finally, the vision statement sets the tone and clearly establishes the company's direction, and it does so without getting into numbers.

HOW TO WIN OTHERS TO YOUR VISION

She is a lonely leader who has no followers. In fact, a leader without followers is no leader at all. Leadership would be a much easier endeavor if the leader had to do nothing more than say to stakeholders, "Follow me." But

---

**Strategies for
Winning Commitment to Your Vision**

- Involve those whose commitment is needed in developing the vision (key stakeholders).
- Establish common ground with stakeholders.
- Listen to stakeholders and consider their input.
- Make the vision personal to stakeholders.

FIGURE 1.4 Followers must share your commitment to the vision.

alas, this is not the case. Technical professionals in official leadership positions (positions of leadership based on authority) can say "follow me" to their direct reports and these subordinates must, of course, give the appearance of following. Truly following and just appearing to do so are vastly different undertakings, however. Those who will truly follow the leader instead of simply making an outward show of doing so are the stakeholders who share the leader's vision and are equally committed to it. Such individuals do not have to be told to "follow me." Rather, the leader is apt to find them prodding her along.

Figure 1.4 shows four strategies for winning the commitment of stakeholders to your vision. These strategies and how technical professionals can use them are explained in the following paragraphs.

Involve Key Stakeholders in the Development of Your Vision

People are more likely to support a vision they played a role in developing. Consequently, it makes sense to involve stakeholders from the outset when developing your vision. In this way the vision becomes *ours* instead of *yours*. It may not be feasible to involve all stakeholders or even representatives of all stakeholder groups in the development of your vision, but it is feasible to involve key stakeholders. For the CEO of a technology company, key stakeholders are her executive-level managers. For a department-level manager, key stakeholders would be other managers and supervisory personnel in the department. For a work team, key stakeholders would be the senior team members (by relative importance of position, not length of service).

Establish Common Ground with Stakeholders

Before beginning development of the vision, take some time to establish common ground with those who will be involved in the process. Find out what drives them, what inspires them, how high you can aim with them, and what they would consider aiming too low. This kind of discussion undertaken before actually developing the vision can make the entire process much easier.

Listen to Stakeholders and Consider Their Input

Involving stakeholders in the development of the vision and then ignoring their input is worse than not involving them at all. This does not mean that all input from stakeholders must eventually find its way into the final vision statement, nor does it mean that their opinions cannot be overruled. The goal is to develop a worthy vision, not to satisfy the egos of everyone involved in developing the vision. It is important, however, for stakeholders to know that their input is listened to, taken seriously, and given full consideration. When they know this, it becomes less important if some of their input is eventually overruled or discarded.

Make the Vision Personal to Stakeholders

This step is critical. There is an adage from politics that says when people go into a voting booth, they vote their wallet, not their conscience. This means that they vote for the candidate who will best serve their personal interests, whether they like that particular politician or not. This personal-interest phenomenon also applies to visions. Stakeholders will commit most readily to the vision that appeals to them on a personal level. In his book *Developing the Leader within You*, leadership expert John C. Maxwell, speaking of stakeholders, says that leaders must "paint the picture for them" and "put things they [the stakeholders] love in the picture."[5]

Maxwell illustrates what he means: "During World War II, parachutes were being constructed by the thousands. From the workers' point of view, the job was tedious. It involved crouching over a sewing machine eight to ten hours a day and stitching endless lengths of colorless fabric. The result was a formless heap of cloth. But everyday the workers were told that each stitch was part of a lifesaving operation. They were asked to think as they sewed that each parachute might be worn by their husbands, their brothers, or their sons."[6]

In other words, those who led the teams of employees doing the monotonous job of sewing parachutes showed workers that they were not just sewing but were saving lives—possibly even the lives of loved ones. This was a brilliant way to make the job personal. This same strategy applies when developing a vision. Employees who can see that they have a personal stake in the vision will more readily commit to it.

Leadership Profile   John Chambers and Cisco Systems

When John Chambers took over as CEO at electronic commerce giant Cisco Systems, he had a vision for where he wanted to take the company. He saw the changes that were taking place in the world of commerce, and he wanted to position Cisco to be at the forefront of that change. Cisco's sales the year Chambers became the company's CEO were $1.2 billion. Within just eight years, Chambers led the company to the $10 billion level. He accomplished this amazing feat by developing a vision, sharing that vision with all of Cisco's employees, and developing farsighted strategies for achieving the vision.[7]

Chambers wanted Cisco to rank either number one or number two in every major segment of the computer-networking market. He also wanted to offer customers one-stop Internet shopping. This was the vision, and he gave all employees business cards with this vision printed on the back. Chambers knew that to accomplish this lofty vision Cisco would have to grow internally while simultaneously acquiring other companies. To make sure that all acquired companies fit in with where he was leading Cisco, Chambers established five ground rules to guide all acquisitions. Two of his five rules dealt directly with the concept of vision: (1) both companies—Cisco and the acquisition—must share a common vision, and (2) the acquisition had to have long-term strategic potential.

The big-picture strategy Chambers adopted to achieve his vision for Cisco was to create unprecedented opportunities for all stakeholders to benefit. Those stakeholders included customers, employees, shareholders, and corporate partners. The beauty of this strategy was that it gave everyone involved incentives by creating opportunities for all stakeholders to succeed to the same extent that Cisco succeeded. By establishing the concept of mutual benefit and by letting every stakeholder know what he was trying to do, Chambers enjoyed unprecedented success, achieving market capitalization of $100 billion in just 12 years—a feat that took software giant Microsoft 20 years to achieve.

IMPEDIMENTS TO DEVELOPING THE VISION

It is important to understand several factors that can inhibit development of the vision. These factors often reveal themselves in the personalities of stakeholders involved in the development process. Leaders should be on guard for stakeholders whose personalities or behaviors can hinder the development of a good vision. Such people should be dealt with to the extent possible during the development process. However, participants whose involvement is detrimental to the process and who do not respond to tactful and patient efforts of the leader should be removed from the development team. People-related factors that can inhibit the development process are as follows:[8]

1. *Thinking too narrowly and too concretely.* People involved in developing a vision must be able and willing to think outside the box and consider the possible. People who cannot think beyond the present realities will hinder the process.

2. *Dogmatically presenting opinions.* Some people present with certainty opinions on subjects that, in reality, they know little about. Such people are also prone to present their opinions as facts and to dare others—through tone of voice, nonverbal cues, and other dogmatic nuances—to disagree. Such people also tend to be confrontational. Dogmatism will not just hinder the development process but also lead to a poorly crafted vision.

3. *Focusing on past failures.* There are those who relish the opportunity to say, "We tried that in the past and it didn't work" and will say it continually. Although corporate memory has its place in the development process, those who relish focusing on past failures bring little or no value to the process.

4. *Being wedded to tradition.* Traditions are an important part of the culture in many organizations. Traditions can be used to help people remember the achievements of those who preceded them, and they should be. In this way the current generation can stand on the shoulders of previous generations and reach much higher than if they ignore tradition and start over again on the ground floor with each successive leadership team. But traditions from the past have value in the present only to the extent that they advance the organization toward a better tomorrow. Consequently, people who are so wedded to tradition that they deny the need for change can hinder the development of a vision that will necessitate change.

5. *Seeking consensus rather than ideas.* Some people are so uncomfortable disagreeing with others or having an opinion that differs from those of others in the group that they seek consensus rather than pursue a worthy vision that is difficult to hammer out. With such people, agreement among participants becomes the goal rather than developing the best possible vision for the organization. Consensus building has a place in the development process, but it comes at the end of the process, not the beginning. Once the development team has established a vision, consensus among other stakeholders who did not participate in the development process will be important. But consensus building is not appropriate until a well-crafted vision has resulted from thorough discussion and vigorous debate, no matter how uncomfortable the debate might make some participants.

6. *Focusing too intently on problems.* Nothing worthwhile comes easily. The higher you set your sights, the more difficulties you will encounter in hitting the target. This is to be expected. One of the main reasons for developing a vision is to give people a beacon in the distance on which they can focus when problems arise, as they invariably will. There are some people who, when developing a vision, can see nothing but the problems that

stand between the reality of today and the possibilities of tomorrow. Such people will be like an anchor around the necks of those trying to develop a vision. Developing a vision is not a pie-in-the-sky endeavor. Problems should be identified and discussed. If they are substantial enough, they might even cause a change in the final vision. But focusing too intently on problems is counterproductive. If an organization waits to move forward until it can adopt a vision that will involve no challenges or generate no problems, it will either never move forward or it will adopt a meaningless vision.

7. *Giving in to the self-interested.*   Some people join the development team to ensure that the vision ultimately adopted serves their personal interests. They are looking for a vision that will help them build their own kingdom, expand their territory, or improve their personal circumstances in some other way. The vision must be developed for the benefit of the organization, not for any self-interested unit or individual.

8. *Giving voice to prophets of doom.*   Some people cannot wait to tell others why their ideas will not work. These people never seem to propose any ideas of their own. Rather, they just make a career out of predicting the certain failure of those proposed by others. Prophets of doom will quickly throw a wet blanket over the development process. Regardless of how important such people are as stakeholders, they should be either kept on a tight rein during the development process or removed from it altogether.

LESSONS ON VISION FROM SELECTED EFFECTIVE LEADERS

As explained in the Introduction, one of the best ways to learn about leadership is by studying the lessons of great leaders from different fields. Following are excerpts from the lives of selected leaders that exemplify some of the vision-related principles set forth in this chapter. The leaders selected for this chapter are Abraham Lincoln, Robert E. Lee, Vince Lombardi, Dorothea Dix, and J. C. R. Licklider.

Abraham Lincoln

Abraham Lincoln was not a man you would pick out of a crowd and say, "Now there goes a man who looks like a leader." In fact, Lincoln may be one of history's best examples of the importance of substance over style. He certainly proved the point that one should never judge a book by its cover, because, in Lincoln's case, the cover was not very attractive. In fact, it was an undeniably homely cover. Tall and gangly with stooped shoulders, Abraham Lincoln walked with the awkward gait of a man with chronically sore feet, and he spoke with a folksy country drawl more suited to a small-town sheriff than an American president. But as he proved time and again, looks can be deceiving, and substance is more important in a leader than either style or image.

Abraham Lincoln was the 16th president of the United States, and he served during the most difficult and challenging period in America's history. "Only ten days before Abraham Lincoln took the oath of office in 1861, the Confederate States of America seceded from the Union taking all Federal agencies, forts, and arsenals within their territory. To make matters worse, Lincoln, who was elected by a minority of the popular vote, was viewed by his *own* advisors as nothing more than a gawky, second-rate country lawyer with no leadership experience."[9]

Vision was one of Lincoln's most enduring strengths as a leader. His career teaches many different lessons about vision, most prominently the following:

Have a vision that is lofty but tangible and capable of appealing to stakeholders on a personal level.

Communicate your vision to stakeholders constantly.

It is well known and documented that during the Civil War Abraham Lincoln, through his speeches, writings, and conversations, "preached a vision" of America that has never been equaled in the course of American history. Lincoln provided exactly what the country needed at that precise moment in time: a clear, concise statement of the direction of the nation and justification for the Union's drastic action in forcing civil war. . . . His vision was simple, and he preached it often. It was patriotic, reverent, filled with integrity, values, and high ideals. And most importantly, it struck a chord with the American people. It was the strongest part of his bond with the *common* people.[10]

So what was Lincoln's vision for America? Simply put, it was the same vision set forth by Thomas Jefferson in the Declaration of Independence. Lincoln envisioned a united country in which all people enjoyed the fruits of liberty, an equal chance, and a government that is of the people, by the people, and for the people. During his time in office, Lincoln never missed an opportunity to press home the main themes of his vision: equality, freedom for all citizens, a fair chance for all, and elevating the condition of mankind.[11]

Robert E. Lee

One might think the juxtaposition of Abraham Lincoln and Robert E. Lee as examples of visionary leaders somewhat odd. After all, Lincoln fought with great passion and skill in the political arena to hold the Union together during America's Civil War, while Robert E. Lee fought with equal passion and skill in the military arena to dissolve it. However, although they fought on opposite sides militarily and philosophically, both Lincoln and Lee were great leaders deserving of study and emulation.

"Robert E. Lee is best remembered as the superb general who held the Union commanders at bay until his outnumbered, outgunned, and poorly fed army was finally overwhelmed by superior forces."[12] Lee's strategies and tactics are studied even to this day in military colleges all over the world, but it is Lee the college president, not Lee the brilliant military commander, who is our example in this case.

Following the Civil War, as president of Washington College (now Washington and Lee University), Robert E. Lee took a small, provincial institution that had been devastated by the war and rebuilt it into a dynamic, competitive university of excellent standing in the world of higher education. His tenure as a college president teaches the following lessons about vision:

> Organizations with a static vision will find it difficult, if not impossible, to compete in a rapidly changing world.

> An organization's vision should be reviewed periodically and updated as necessary to accommodate ever-changing conditions.

When Lee took over as president of Washington College, things had been the same for many years, and, even without the devastation wrought by the Civil War, the college was sitting still while the world passed it by. Unfortunately, "the way we have always done things" was the way Lee's faculty and staff wanted to continue doing things. The college had only one curriculum—the classics—and all students took the same courses. Because of his many battles with a better-equipped, more technologically advanced enemy during the Civil War, Lee saw clearly that the old agrarian ways of the South would have to change if that war-torn region was to recover from the scorched earth policy of the North's victorious army and be a player in an increasingly technological world. In short, Lee knew that having students study only the classics would not put food on Southern tables, rebuild the South's physical and economic infrastructure, or help spawn badly needed economic development.

Consequently, Lee put in place a strategy that for the time was both controversial and risky. He retained the classics as the basic academic core for all degrees, but he added new courses and programs in such practical fields as business, agriculture, and engineering. He also established a fellowship program for promising graduate students, a scholarship to support students majoring in journalism, and summer studies.[13] Although these innovations probably seem tame by today's standards, in the post–Civil War South no college president of lesser stature than Lee could have survived even proposing them.

Vince Lombardi

"Legendary coach Vince Lombardi—loved by some, feared by others, respected by all—was first and foremost a winner. His unparalleled ability to inspire greatness and mold disparate groups of individuals into dominating championship teams made Lombardi an icon both on and off the playing field."[14] "As coach of the Green Bay Packers from 1959 to 1967, Coach Lombardi turned a perennial loser into a juggernaut, winning NFL titles in 1961, 1962, and 1965 in addition to Super Bowls I and II in 1966 and 1967. Lombardi's uncanny ability to motivate his players, his extraordinary capacity to inspire them to test and exceed the limits of their physical and mental endurance, and his insatiable drive to succeed made him the standard against which all NFL coaches are measured."[15]

Lombardi took over the Green Bay Packers in 1959. Within just two years he had turned the team into a winner and realized his vision of an NFL championship, but he did not stop there. Instead of being satisfied with his laudable success, Lombardi recast his vision, aiming even higher. His next vision was to win championships consistently, something only a few NFL teams have ever achieved. Finally, Lombardi set his sights on winning what came to be known as the Super Bowl. The vision lessons Vince Lombardi's career as an NFL coach teach are many. Three of the most important and compelling are as follows:

Your vision must be meaningful to those you would lead.

Your vision must inspire those you would lead.

Never be satisfied with your success—recast your vision as performance and progress allow, setting your sights ever higher.

When Lombardi took over as coach of the Green Bay Packers, the team was a perennial loser, not just in terms of games won and lost but also in terms of its self-image and attitude. The players looked, played, and thought like losers. The vision Lombardi shared with his new team was simple: The Green Bay Packers would be NFL champions. After communicating this simple but lofty vision to them, he spent much time and energy helping his players see how achieving this vision would help improve their financial prospects, enhance their self-respect, and win the esteem of their families, friends, and fans. This made the difficult challenge he was asking them to commit to achieving meaningful. The player who did not give his best on a Vince Lombardi team was made to feel as if he was doing nothing less than letting his family down, and he was—both his personal family and his team family. Winning an NFL championship was not just meaningful to Lombardi's players but also inspirational. People who work hard enough to become professional football players are competitive by nature. They want to win. Consequently, they responded to Lombardi's vision on a deeply personal level.

Dorothea Dix

"Dorothea Lynde Dix was a pioneer and leader in the humanitarian reform of the treatment afforded convicts and the institutionalized mentally ill."[16]

"She also led a groundbreaking international crusade for the rights of prisoners and the mentally ill. Born in Hampden, Maine, Dix left her schoolteacher's position in 1836 to travel abroad for reasons of health. While
visiting England, she encountered the British prison reform advocate Henry
Tuke, whose ideas she found immensely inspirational. She returned to the
United States in 1841, determined to work toward the improvement of
prison conditions."[17]

The lessons about vision taught by the career and crusade of Dorothea
Dix are many. Prominent among these lessons are the following:

One person committed to a lofty vision can make a difference.

A vision must be communicated to stakeholders in ways that make them
want to do the right thing.

Dix's vision was to ensure the proper and humane treatment of prisoners, especially those who were mentally ill. She began pursuing her vision
by accepting a position as a Sunday school teacher at a jail in Massachusetts.
Although this was a modest beginning for a person who would eventually
achieve unprecedented results, the time Dix spent in this position affirmed
that she had found her calling in life. It soon became apparent to Dix that
people suffering from mental illnesses were being warehoused in prisons
and that the prison environment only served to worsen their conditions.

Dix began her crusade for better conditions and more humane treatment of the mentally ill with a letter-writing campaign. She crafted
conscience-tugging letters to newspapers that appealed to readers on a deeply
emotional and personal level. She also drafted reports for state legislatures
and petitioned the U.S. Congress to authorize funding for the treatment of
the mentally ill. Dix took every opportunity to communicate her vision to
anyone and everyone who might be able to help her achieve it. Eventually,
her efforts bore fruit. The crusade carried out almost single-handedly by Dix
resulted in passage of reform legislation in several states, modernization of
facilities in other states, construction of more than 30 mental hospitals, and
development of a model for the humane and therapeutic treatment of mentally ill prisoners that was used not just in the United States, but also in
Canada, Russia, and Turkey. Armed with nothing more than a lofty vision
and ink pen, Dix became a leader who made a difference.

Leadership Tip

"It's amazing what ordinary people can do if they set out without preconceived notions."[18]

—Charles F. Kettering
American Engineer and Inventor

J. C. R. Licklider

Those of us who use the Internet every day have a man with the unlikely name of J. C. R. Licklider to thank. Actually, we have numerous other leaders from the fields of government, the military, technology, and business to thank as well as Licklider, because the Internet's development was not a one-man show. However, as is always the case, whether building a house or installing a new technological innovation, work begins with the foundation, and the visionary foundation of today's Internet was established by Licklider in 1963.[19]

Licklider was the head of computer research for the Pentagon's Advanced Research Projects Agency (ARPA), an organization created in 1958 by President Dwight Eisenhower in response to the Soviet Union's successful launch of the Sputnik satellite. Americans were embarrassed and a little frightened to have the Soviets push ahead of the United States in space technology, or any other technology for that matter. Eisenhower and other governmental and military leaders were determined to ensure that the Soviet Union's lead in space would be temporary and short-lived.

Licklider's foundational contribution to the development of the Internet was to offer the first coherent vision for a universal computer network. Other leaders from various fields then took his vision and did their part to make it a reality. The vision lesson taught by the career of J. C. R. Licklider is

Leaders seldom achieve anything of significance by themselves.

In 1967 Larry Roberts was hired by ARPA to transform Licklider's vision into reality. He responded by proposing electronic linkages among a number of institutes via what would come to be known as ARPANET. In 1969, ARPANET established nodes at UCLA, Stanford Research Institute, the University of California at Santa Barbara, and the University of Utah. In 1971 Ray Tomlinson, a private businessman, invented email and adopted the "@" symbol. The first commercial version of ARPANET, known as Telenet, was established in 1974. In 1976 Robert Metcalf of Xerox's Palo Alto Research Center demonstrated the first local area network, known as the Ethernet. By 1977 the number of Internet host computers exceeded 100.

The Internet as known today was established in 1982 when ARPA introduced a standard networking language—TCP/IP—developed primarily by Robert Kahn and Vinton Cerf. A major leap forward occurred in 1986 when the National Science Foundation established NSFNET, which connected five supercomputing centers. This touched off a tidal wave of new connections—many in major universities. In 1991, the European physics laboratory known as CERN released the World Wide Web, developed by Tim Berners, as a vehicle for sharing information in the physics community. In 1993, the first Web browser, known as MOSAIC, was introduced at the University of Illinois by Mar Andreesen. Today the Internet connects almost 700 million people worldwide and almost 200 million Internet hosts.

J. C. R. Licklider probably never imagined where his vision would lead, but that is the beauty of a well-developed vision. It can take an organization, an idea, or a technology beyond even the realm of its author's dream. Licklider is just one in a long list of leaders of technology who are responsible for the Internet, but he is the one who had the vision that got all of the others started.

Summary

1. A vision is a picture of the leader's intended destination for the organization she hopes to lead.

2. A good vision has the following characteristics: easily understood by all stakeholders; briefly stated, yet clear and comprehensive in meaning; challenging, yet attainable; lofty, yet tangible; capable of winning the commitment of stakeholders; capable of setting the tone for stakeholders; and able to establish direction without getting into numbers.

3. A vision is important because it gives meaning, direction, focus, and inspiration to the workers in an organization.

4. A vision is developed by studying visionary leaders, considering the possible, establishing a context, examining the present, studying the future, identifying the ideal condition within a given context, asking the legacy question, summarizing your ideas in a concise, written vision statement, and testing this statement against the characteristics of a good vision.

5. The following strategies will help win others to your vision: involve key stakeholders in the development of your vision, establish common ground with stakeholders, listen to stakeholders and consider their input, and make the vision personal.

6. Factors that can hinder development of the vision are as follows: thinking too narrowly and too concretely, dogmatically presenting opinions, focusing on past failures, being wedded to tradition, seeking consensus rather than ideas, focusing too intently on problems, giving in to the self-interested, and giving voice to prophets of doom.

7. Two vision lessons from Abraham Lincoln are (a) have a vision that is lofty but tangible and capable of appealing to stakeholders on a personal level; and (b) communicate your vision constantly.

8. Two vision lessons from Robert E. Lee are (a) organizations with a static vision will find it difficult, if not impossible, to compete in a rapidly changing world; and (b) an organization's vision should be reviewed periodically and updated as necessary to accommodate ever-changing conditions.

9. Three lessons about vision from Vince Lombardi are (a) your vision must be meaningful to those you lead; (b) your vision must inspire those

you lead; and (c) never be satisfied with success—recast your vision as progress allows, setting your sights ever higher.

10. Two vision lessons from Dorothea Dix are (a) one person committed to a lofty vision can make a difference; and (b) a vision must be communicated to stakeholders in ways that make them want to do the right thing.

11. The vision-related lesson of J. C. R. Licklider is that leaders accomplish little alone.

Key Terms and Concepts

Able to establish direction

Ask the legacy question

Being wedded to tradition

Briefly stated, yet clear and comprehensive

Capable of setting the tone

Capable of winning commitment

Challenging, yet attainable

Dogmatically presenting opinions

Easily understood by stakeholders

Establish common ground with stakeholders

Establish your context

Examine the present

Focusing on past failures

Focusing too intently on problems

Giving in to the self-interested

Giving voice to prophets of doom

Identify the ideal condition within your context

Involve key stakeholders

Listen to stakeholders and consider their input

Lofty, yet tangible

Make the vision personal

Seeking consensus rather than ideas

Study the future

Study visionary leaders

Test the vision statement

Thinking too narrowly and too concretely

Vision

Write your vision statement

Review Questions

1. Discuss the concept of vision.

2. Explain why it is important to have a clear vision.

3. Explain how to go about developing a vision.

4. Explain how you would win the commitment of others to your vision.

5. List and explain the various factors that can inhibit the development of a vision.

6. Explain the lessons about vision taught by Abraham Lincoln's career and give examples.

7. Explain the vision lessons taught by the second career of Robert E. Lee as a college president and give examples.

8. Explain the vision lessons taught by the career of Vince Lombardi and give examples.

9. Explain the lessons about vision taught by the crusade of Dorothea Dix and give examples.

10. Explain the vision lessons taught by the career of J. C. R. Licklider and give examples.

LEADERSHIP SIMULATION CASES

The following simulations are provided to generate additional thought and discussion about the principles of leadership explained in this chapter. Readers are encouraged to consider how the situations presented in these cases might apply to them and to discuss the cases with other leaders and leadership candidates.

CASE 1.1 Why Do I Need a Vision Anyway?

Tom Prince has just been promoted to his first leadership position with ABC Engineering and Technology Company, Inc. ABC is a medium-size government contractor that specializes in retrofitting and modernizing fixed-wing military aircraft. The department Prince will manage for ABC is mechanical engineering. Prince is an excellent engineer with more than 10 years of experience. However, his career to this point has consisted of various design and trouble-shooting positions that involved no supervisory or leadership responsibilities. His new position is the first he has had that will require leadership skills. Aware of Prince's limited experience, ABC's vice-president for engineering required him to enroll in a leadership training program that Prince has just started. In the first lesson, the instructor is covering the need for a vision and how to develop it.

Tom Prince and several of his fellow leadership students are having lunch together and discussing what they learned during the morning session. Prince, who is not enjoying the training, asks his colleagues, "Why do we need a vision anyway? Employees have job descriptions; those of us who are managers give them their assignments and tell them when the work is due. I don't see what good a vision is going to do." One of his colleagues who appeared to feel the same way responded, "Don't ask me Tom. This is my first leadership seminar too."

Discussion Questions

1. How would you explain the need for a vision if you were one of the people sitting at the table having lunch with Tom Prince?
2. What kinds of problems might Tom Prince have managing his new department if he fails to develop a good vision for the department?

CASE 1.2 First Attempt at Writing a Vision Statement

Sherry Anderson had more than 15 years of software development experience when the company she worked for was bought out in a hostile takeover. When the dust had settled from the takeover, Anderson found herself with a substantial severance package but no job. Undeterred, she decided to view her new circumstances as an opportunity to do something she had been mulling over for years. Using the money from her severance package, Anderson started her own company—Anderson Software, Inc.

Anderson had worked as a high school computer instructor for five years before going into the private sector. Drawing on this background, she decided to build her new company by developing instructional software for applications at the high school and college levels. Her company has been in business for a little more than a year, and things are looking up. Now that she is able to focus on something other than making payroll and paying the rent, Anderson is trying to develop a vision for her company. She and her top managers have developed a first draft, but they are not comfortable with it. The draft reads as follows:

Anderson Software, Inc., will be the best in its market and will grow by at least 10 percent a year for the next five years.

Discussion Questions

1. Without referring to the characteristics of a good vision, what is your analysis of this vision statement?
2. Now, using the characteristics of a good vision statement, conduct a more thorough analysis of the statement and make recommendations as if you were making them to Sherry Anderson.

CASE 1.3 How Can I Get My Employees to Commit to the Vision?

In simulation Case 1.2, Sherry Anderson and her top managers were trying to develop a vision for Anderson Software, Inc. Assume they were able to develop a good vision statement that has all of the characteristics of a good vision. Anderson now wants to gain the commitment of her employees to the vision. Speaking to her top managers, Anderson asks, "Now that we have a good vision, how should we go about convincing our employees to share our commitment to it?"

Discussion Questions

1. Play the role of a top manager with Anderson Software, Inc. How would you answer Sherry Anderson's question?

2. What problems might Anderson encounter as she tries to win the commitment of her employees? How would you recommend these problems be dealt with?

Endnotes

[1] Louis E. Boone, *Quotable Business*, 2nd ed. (New York: Random House, 1999), 8.

[2] James M. Kouzes and Barry Z. Posner, *The Leadership Challenge*, 3rd ed. (San Francisco: Jossey-Bass, 2002), 112.

[3] Louis E. Boone, 8.

[4] Donald T. Phillips, *Lincoln on Leadership* (New York: Warner Books, 1992), 167–168.

[5] John C. Maxwell, *Developing the Leader Within You* (Nashville, TN: Thomas Nelson, 1993), 154–156.

[6] John C. Maxwell, 156–157.

[7] Thomas J. Neff and James M. Citrin, *Lessons from the Top* (New York: Currency/Doubleday, 1999), 79–81.

[8] John C. Maxwell, 150–154.

[9] Donald T. Phillips, back cover.

[10] Donald T. Phillips, 163.

[11] Donald T. Phillips, 165.

[12] Al Kaltman, *The Genius of Robert E. Lee* (Upper Saddle River, NJ: Prentice Hall, 2000), 328–329.

[13] Al Kaltman, 328.

[14] Vince Lombardi Jr., *What It Takes to Be # 1: Vince Lombardi on Leadership* (New York: McGraw-Hill, 2003), 145–146.

[15] Vince Lombardi Jr., inside front cover.

[16] Alan Axelrod, *Profiles in Leadership* (Upper Saddle River, NJ: Prentice Hall, 2003), 145–146.

[17] Alan Axelrod, 145.

[18] Louis E. Boone, 8.

[19] David Lagesse, "Speeding the Net," *U.S. News & World Report*, Special Collector's Edition, American Ingenuity, 2002, 52–55.

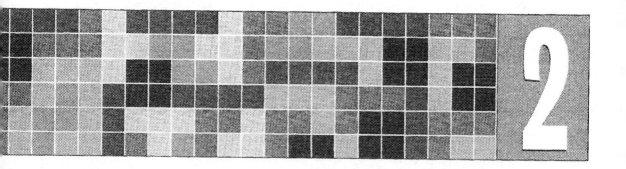

Project Unquestionable Integrity and Selflessness

"Image is based on how we look and what we do. Integrity is based on our character—who we really are. In the long run, people will see through our image to our character—good or bad."

—The Author

OBJECTIVES

- Define the concept of integrity as it applies to leadership.
- Define the concept of selflessness as it relates to integrity.
- Explain why integrity and selflessness are so important to those who would lead.
- Explain the ethical dimension of leadership.
- Complete a personal self-assessment on integrity.
- Summarize the integrity-related lessons of selected effective leaders.

Leadership is about character more than anything else, and integrity is the most important aspect of character to those who would lead. Just how important is integrity? Consider the following story.

An old man decided that he would return to his native country of Silesia before becoming too infirm to make the trip—a trip that would require him to travel on horseback and on foot through a dangerous forested region of Poland. This region was known to be inhabited by cutthroat bandits.[1] As the old man traveled the heavily forested path that would eventually lead him home, he kept a sharp eye out for the thieves reputed to be lurking there. In spite of his vigilance, a band of stealthy robbers was able to surprise and accost him. They demanded all of his valuables, which he promptly but shakily handed over. The robbers demanded to know several times if the old man had given them everything. After assuring his assailants repeatedly that he had, they let him be on his way. Relieved just to have lived through the incident, he walked as fast as his old legs would carry him, trying to put as much distance as possible between himself and the bandits.

However, before making much progress, he felt something hard banging against his shin as he walked. Only then did the old man remember that he had sewn a gold piece into the hem of his robe for safekeeping. He immediately retrieved the gold piece and hurried back down the path to catch up with the bandits who had robbed him. When he finally overtook them, the old man told their leader that he had unintentionally told them a lie. Handing over the gold coin, the old man explained that he had not intended to deceive them, but had simply forgotten about the coin hidden in the hem of his robe. To the old man's astonishment, the bandits not only refused to take the gold coin but also gave back all of his valuables and helped him on his way.[2]

The old man's strength of character and integrity not only restored his valuables but also made a deep impression on a group of bandits. This story is a powerful example of the positive effect integrity can have on people— even those who may lack it themselves.

INTEGRITY DEFINED

Integrity is strict adherence to a code of ethics that a person has internalized. A person with integrity has adopted a positive value system, knows what beliefs make up that system, and lives according to those beliefs regardless of the circumstances, peer pressure, present company, or other influences. A person with integrity will make the same choice and behave the same way whether in a crowd or alone and whether observed or unseen. People with integrity have in common the following traits:

1. *Refusal to pretend.* There is no hypocrisy in people with integrity, nor do they flow with the ever-changing currents of human behavior or allow peer pressure to determine their decisions.

2. *Internal moral compass.*   Their values are so much a part of them that the person and the values cannot be separated. All aspects of their daily lives are guided by their internal moral compass.

3. *Consistency in behavior.*   Their everyday behavior is guided by their internal moral compass. Consequently, it is consistent and predictable. Said another way, they are the same person when they are out of town as when they are at home.

4. *Consistency in decisions.*   Their decisions are based on what is best for the organizations they lead—within the context of ethical behavior—rather than misguided self-interest.

5. *More concern for substance than image.*   They know that even the most carefully crafted image will not stand up to the day-to-day pressures imposed on leaders in the workplace. Consequently, they focus on substance, knowing that in the long run integrity will win out over an artificial, surface-level image.

6. *Selflessness.*   One cannot be a self-centered, self-serving person when dealing with others and also be a person of integrity. Leaders must be good stewards of the resources for which they are responsible—human, financial, and physical. This means the leader must think first of the organization and people he hopes to lead and put their needs—in the context of the organization's goals—ahead of his own.

SELFLESSNESS AND INTEGRITY

Selflessness is the opposite of that all-too-pervasive philosophy of life "I'm looking out for Number 1." Selflessness is an inseparable by-product of integrity. People with integrity invariably are selfless, because acts of integrity are by their very nature selfless acts. When an individual risks his personal interests to do the right thing, as opposed to the popular thing, he is performing a selfless act. The two concepts are inseparable when used in the context of leadership. The following example shows the positive effect selflessness can have on an individual's ability to lead.

John Martin had been an engineer for Alpha-Beta, Inc. (ABI), for more than 15 years when an opportunity for leadership was unexpectedly thrust upon him. After several months of difficult negotiations, ABI finally beat out 12 other competitors to win what those involved had come to call the Mega-Stadium Project. The Mega-Stadium, when completed, would be the largest sports complex in the United States, containing a professional football stadium, basketball arena, baseball park, and soccer field.

The good news was that ABI won the contract. The project would ensure the company at least five years of engineering work and could give ABI an international reputation, which would surely lead to future business. The bad news was that ABI had never attempted a project of this magnitude, and, to complicate matters, the company's chief engineer—the person who had masterminded ABI's successful proposal and was slated to lead the project—died unexpectedly the day after the contract was awarded. This sad fact was the reason that responsibility for leading ABI's Mega-Stadium Project had fallen on John Martin's shoulders.

At the time of his elevation to manager of the Mega-Stadium Project, John Martin was ABI's most capable civil and structural engineer. He had trained and mentored most of the technical personnel assigned to the project, and several of ABI's most noteworthy projects were his designs. However, Martin's responsibilities had always been in the area of design, not project management or leadership.

Over the next several years as he managed the Mega-Stadium Project, John Martin learned a great deal about leadership, and he did it the hard way. He made mistakes, but he learned from them and moved on. There were plenty of rough spots on the road from project initiation to project completion, but Martin and his team persevered and completed the engineering work for the Mega-Stadium on time.

More than any other characteristic, John Martin's selflessness allowed him to hold his team members together and convince them to work the many nights, weekends, and holidays that turned out to be necessary to complete the Mega-Stadium Project on time. For 15 years before being assigned to lead this huge, career-altering project, Martin had often devoted his free time to tutor, train, and mentor other employees at ABI. He had helped them catch up when they fell behind in their work. He had spent countless evenings and weekends standing in for colleagues and subordinates when the workload had necessitated overtime, so that they could honor family obligations too important to miss. Because of Martin's long record of selflessness, the employees assigned to the Mega-Stadium Project were willing to stick with him and give him the chance to grow and learn as their leader. Because they knew he never asked them to work late unless he planned to work even later, Martin's team members were willing to follow his example and work the long hours that were necessary. The selflessness of its project leader, more than any other factor, was the foundation on which the Mega-Stadium was built.

IMPORTANCE OF INTEGRITY AND SELFLESSNESS TO LEADERS

A leader without followers is no leader at all. This is why integrity and selflessness are so important to those who would lead. A leader can lead only if people will follow, and people will follow only if they believe in, trust, and respect the leader. Belief, trust, and respect are all founded in the integrity and selflessness of the leader. Without these characteristics, a leader will be without followers. In his book *Developing the Leader within You*, John Maxwell gives the following reasons why integrity is so important:[3]

1. *Integrity builds trust.* People will not just follow leaders they trust but will go the extra mile and commit to helping them succeed. Those they distrust they will only reluctantly and halfheartedly "go along with." Leaders who are trusted get action from their followers. Leaders who are not trusted get lip service, halfhearted compliance, and tactical disobedience (someone pretending to comply who, in fact, is not).

2. *Integrity leads to influence.* To lead people, you must be able to influence them sufficiently to gain their confidence and commitment. People are not influenced by those they cannot trust. Leaders who fail to win the trust of their followers will not have the influence necessary to achieve buy-in, commitment, and willingness to persevere through the inevitable obstacles that block the pathway to success.

3. *Integrity establishes high standards.* Leaders with integrity consistently set a positive example of doing the right thing regarding the organization's goals, as opposed to taking a more expeditious or self-serving path. Their actions convey the message "Do as I do" rather than "Do as I say, not as I do." By their positive examples, leaders set high standards for their followers, and high standards are necessary for success in the global marketplace.

4. *Integrity establishes a solid foundation rather than just an image.* Many people who want to be leaders spend a great deal of time, money, and effort trying to look like a leader. They "dress for success," hire personal trainers, spend hours in the gym, and enroll themselves in various image-building seminars. Although image can be a factor in success, image without substance will quickly lead to failure. People who lack integrity will eventually show it regardless of how hard they work to maintain a leadership image. For those who would lead, a reputation for integrity must come first, and image second. The good news is that, in the long run, if you establish a reputation for integrity, your image will take care of itself.

5. *Integrity builds credibility.* Credibility is what you have when others believe in you, and it is a critical characteristic for those who would lead. To establish credibility, there must first be trust. People will not believe in someone they do not trust. If you want to establish credibility, first establish trust.

The excellent research of James Kouzes and Barry Posner into what people look for and admire in their leaders shows how important integrity is to effective leadership.[4] Kouzes and Posner have studied the question of what people look for in their leaders for more than 2 decades, and their research to date has involved more than 75,000 respondents from Australia, Canada, Japan, Korea, Malaysia, Mexico, New Zealand, Scandinavia, and Singapore as well as the United States. Through this research, Kouzes and Posner have identified and ranked 20 characteristics that people want to see in those who would lead them.

Leadership Tip

"I have found that being honest is the best technique I can use. Right up front, tell people what you are trying to accomplish and what you are willing to sacrifice to accomplish it."[5]

—Lee Iacocca
Former Chairman, Chrysler Corporation

For 2 decades, every time they have conducted their survey, the most wanted characteristic in a leader has been honesty (another way to say *integrity*). This holds as true in other countries around the globe as it does in the United States. Honesty is also the highest ranked of the 20 characteristics in every country involved in the survey. Clearly, the importance of integrity in those who would lead cannot be overstated.

THE ETHICAL DIMENSION OF LEADERSHIP

Just as technology is applied science, ethics is applied morality. It follows then that *ethical behavior* means doing the right thing as *right* is defined within a given value system. To behave ethically, a leader must have an established moral framework—a set of internalized values that guide her behavior, decisions, and everyday interaction with other people. To lead well, you must set a consistent and positive example of ethical behavior, and to set such an example, you must have an internal moral compass that guides you.

Leaders are continually faced with ethical dilemmas in which they must balance the needs of individuals and the needs of the organization, make decisions that can affect the organization's profitability in both the short and the long run, balance their personal responsibilities to family and work, and withstand both overt and covert peer pressure to cut corners in the name of short-term profitability. Making decisions that have high ethical content can cause leaders to undergo what is commonly known as soul-searching. This typically amounts to weighing what the leader truly believes is right against the various pressures applied to her during the decision-making process, while also considering the potential personal consequences of deciding one way or the other.

Making unpopular decisions is one of the most difficult and most frequently faced aspects of being a leader, because to make a decision that is right, as opposed to popular, you must have the moral courage and personal resolve to persevere against the inevitable pressure that will come from a variety of sources (e.g., the marketplace, supervisors, subordinates, and sometimes even family members). But the most difficult foe of the leader who wants to do the right thing instead of the popular or expedient thing is that most seductive and tenacious enemy of integrity—the leader's own self-interest. Human beings are driven by self-interest, such as the desire for safety, security (both personal and economic), relationships, the esteem of peers, and so on. Every time a leader makes an unpopular decision, she puts one or more of these interests at risk. Even when they know she is right, people rarely thank a leader for making a decision that runs counter to their desires, because, like the leader, they too are driven by self-interest.

Consequently, leaders should never be so naïve as to expect those opposed to their decisions to just sit back and willingly accept their fate. If the person who is disgruntled about the leader's decision is a customer, he might retaliate by taking his business elsewhere, pointedly blaming this unwelcome action on the leader. The leader then finds his job security (economic self-interest) threatened because his ethically correct decision caused his company to lose an important customer.

If the disgruntled person is the leader's supervisor, the leader who did the right thing might be risking that raise she has worked so hard to earn, or even future promotions. If the disgruntled person is a colleague, the leader's moral stand might cause her to be isolated from her peer group—shunned personally and professionally. If the disgruntled people are direct reports, the leader who did the right thing might find the camaraderie he has worked so hard to instill in his team being replaced by an attitude of reluctant compliance or even tactical disobedience—at least temporarily.

What these few examples say collectively is that leaders sometimes pay a price in personal consequences for doing the right thing. At the very least, the threat is always there. This fact alone prevents many people who would otherwise be effective leaders from ever rising to that level.

If the pressure on leaders to decide in one's favor came only from self-interested stakeholders, effective leadership would be only half as difficult an undertaking as it is. However, even while receiving short-term pressure from stakeholders in the decision-making process, leaders are also pressured by the potential long-term effects of their decisions. Leaders who know what is right but succumb to the predictable pressures of the moment do not have the luxury of saying, "I told you so" when the long-term consequences of their decisions turn out adversely for stakeholders. This is perhaps the most bitter irony universally faced by leaders regardless of their field of endeavor. Those who for reasons of self-interest pressure the leader most persistently to take option A will be first in line to criticize when the long-term results show that the leader—in spite of their pressure—should have taken option B. Newspapers and business journals are replete with stories that chronicle the startling downfalls of companies whose leaders let themselves be pressured into making unethical decisions that served the short-term personal interests of some but had a devastating effect on the company in the long run.

This "double jeopardy" faced by leaders might seem unfair to some, and it probably is. But, then, this is just one of the many reasons why so few people ever become effective leaders or even choose to pursue leadership positions. No person comes to understand the sometimes harsh unfairness of life better than those who lead. When leaders are right, they are heroes, and people are quick to forget the self-interested criticism they aimed at the leader for making what they—the critics—thought was the wrong decision. But when leaders are wrong, they are scapegoats, and circumstances, justifications, and logic rarely matter to their detractors.

We see this harsh fact of life for leaders acted out every year in college football programs. Consider the college football coach whose team loses a major bowl game because he did the right thing and benched a player who had failed to attend classes regularly. Alumni will no doubt call for the coach's lynching at sundown, and in the years to come they will remember only that their team lost the game. Alternatively, assume the coach ignores the eligibility rules, plays the truant team member, and wins the game. Then, two weeks later, after being tipped off by an investigative reporter from the local newspaper, authorities strip the coach and his team of their win because the coach played an ineligible team member. Once again, stakeholders will call for the coach's head. This is the double jeopardy of leadership.

Leadership Profile   Charles Wang and the Value of Integrity

Charles Wang's founding of Computer Associates came about because his grades in college weren't good enough to get him into graduate school. Consequently, instead of pursuing a graduate degree, Wang did what most college graduates do every year—he went looking for a job. The job he eventually found was in the field of computer programming, a fortuitous turn of events for Wang and for the many clients who now depend on business applications software from Computer Associates.

Shortly after he began work as a computer programmer, Wang noticed that the software his employer developed was missing the mark in addressing the needs of customers. Although the software was up-to-date and well developed, it just didn't do what customers needed it to do. After this observation, Wang decided to start his own company. Today Computer Associates ranks at the top of the list of software developers, along with IBM and Microsoft. Nineteen out of the top 20 on the Fortune 500 list use at least one of the software packages developed by Wang's company.

The keys to Wang's success are innovation, customer focus, smart acquisitions, and integrity. According to Wang, a leader cannot be successful in the long term without integrity. He thinks that leaders in business must have a moral compass so that their word can be trusted by those they deal with. According to Wang, lack of integrity will eventually catch up with a business leader and undermine his ability to lead.[6]

MAKING ETHICAL DECISIONS

Leaders sometimes struggle with deciding exactly what is right to do in a given situation. In some cases the struggle is the result of a weak leader's attempts to rationalize a decision he knows is unethical but wants to make anyway for reasons of self-interest. However, there are times when even good, ethical leaders are stumped—they simply do not know the right thing to do in a given situation. Figure 2.1 shows five tests leaders can use to decide what is ethically right before making a decision, as follows.

- *Morning-after test.* It is not uncommon for a person to do something under the pressure of the moment only to wake up the next morning full of regret. Leaders can use this human characteristic to their benefit when making decisions. Before making a potentially controversial decision, ask yourself, "If I make this choice, how will I feel about myself in the morning?"

- *Front-page test.* Before making a potentially controversial decision, ask yourself, "How would I feel if all the details of my decision were printed on the front page of my local newspaper?"

- *Mirror test.* Before making a potentially controversial decision, ask yourself, "If I make this choice, how will I feel about myself when looking in a mirror?"

- *Role-reversal test.* Before making a potentially controversial decision, trade places with the people who will be affected by your decision and view the results through their eyes. How would you feel if you were in their shoes?

- *Commonsense test.* Listen to what your instincts and common sense are telling you. If it seems wrong, it probably is.

Tests for Making Ethical Choices

- Morning-after test
- Front-page test
- Mirror test
- Role-reversal test
- Commonsense test

FIGURE 2.1   Tests leaders can use when making decisions.

ASSESSING YOUR PERSONAL INTEGRITY

We never really know what we will do or how we will react in a new and unfamiliar situation until we actually face that situation. However, a self-assessment of your personal integrity will allow you to at least identify areas that might need improvement. Figure 2.2 is a self-assessment instrument the reader can use to identify personal strengths and weaknesses relating to integrity.

Ideally you would be able to honestly place an A (for always) in the blank preceding each of the 10 statements contained in the self-assessment instrument. However, it is more likely that you will have to place a U (for usually), an O (for occasionally), or an N (for never) in some of the blanks. If this is the case, any item with an O or an N beside it represents an aspect of

---

Self-Assessment Instrument
for Integrity

Rate your personal integrity by placing the appropriate letter in the blank next to each statement below.

N = Never
O = Occasionally
U = Usually
A = Always

_____ 1. People can accept my word as my bond.
_____ 2. I am willing to admit that I am wrong.
_____ 3. I tell the truth in all situations.
_____ 4. People can trust that I will fulfill my promises.
_____ 5. My standards of conduct do not change when I go out of town.
_____ 6. I am selfless in my dealings with others.
_____ 7. I want to be held accountable for my actions.
_____ 8. I try to re-earn the trust of others every day.
_____ 9. I refuse to rationalize as a way to justify a self-serving decision.
_____ 10. I refuse to stretch or withhold the truth to serve my self-interests.

FIGURE 2.2   The ideal response to every question is A, meaning *always*.

Leadership Tip

"The ability to adapt and adjust tactics while sticking to principles is extremely important. One of the biggest problems with CEOs is that they are flexible on principle and inflexible on plans."[7]

—Eugene E. Jennings
American Business Writer

---

integrity you need to work on. Do not be concerned if you cannot honestly record a perfect score; few people can. Honestly admitting you need to improve in certain areas is, in itself, a mark of integrity. You will find that becoming a person of impeccable integrity is more of a journey than a destination.

No human being is likely to ever achieve perfect integrity. It is important to understand this fact or you might become overwhelmed by the magnitude of the challenge. But, neither can you afford to let "nobody's perfect" become an excuse to stop striving for perfection. Although it is true that you are not likely to achieve perfect integrity, it is equally true that the more persistently you try, the closer you will come, and the closer you come, the better you will be at leading others. When you find yourself struggling with ethical questions and the pressure of the moment causes you to lean toward an expeditious decision, remember the following unalterable truth:

There is no right way to do a wrong thing.

LESSONS ON INTEGRITY AND SELFLESSNESS FROM SELECTED LEADERS

Following are excerpts from the lives of selected leaders whose careers exemplify some of the integrity and selflessness principles set forth in this chapter. The leaders selected are Lucius Quintus Cincinnatus Lamar, Clara Barton, Ulysses S. Grant, John Quincy Adams, and Bill Gates.

Lucius Lamar[8]

His full name was Lucius Quintus Cincinnatus Lamar, an appellation derived from that of Lucius Quintus Cincinnatus, the Roman farmer who rose from obscurity to lead the army that rescued Rome from Aequi invaders in 458 B.C. Having won this great victory, Cincinnatus was offered power, wealth, and the opportunity to be dictator of Rome. To the amazement of most, he refused all offers. This simple farmer who could have been king was content to do his duty and then return to a life of obscurity.

Lucius Quintus Cincinnatus Lamar's was an illustrious name that he honored throughout his career as an American political leader. He is remembered not for his legislative agenda or for the bills he sponsored but for the moral courage he showed so many times while bearing the wrath of his parochially minded constituents in order to do what was right for the entire country. Lamar's thoughts on doing what is right instead of what is popular are summed up by a statement he made in 1878 while enduring one of the most pernicious political attacks leveled against him during a controversial career replete with such attacks. He said, "The liberty of this country and its great interests will never be secure if its public men become mere menials to do the biddings of their constituents instead of being representatives in the true sense of the word, looking to the lasting prosperity and future interests of the whole country."[9] The lesson on integrity taught by the career of Lucius Lamar is this:

> A leader must do what is right, not what is popular, even when the
> personal cost is high.

If people are influenced at all by family lineage, then Lucius Lamar was a man destined from birth for greatness. Of French Huguenot descent, Lamar's family tree included such illustrious names in the history of America as Jefferson Jackson, Thomas Randolph, Lavoisier LeGrand, and Mirabeau Buonaparte. Buonaparte, who was Lamar's uncle, led the charge at San Jacinto that broke the Mexican line, an act of heroism that propelled him to even greater heights as the second president of the Republic of Texas.

While working as an attorney in Oxford and teaching at the University of Mississippi, Lamar's rabid pro-Southern point of view brought him to the attention of local power brokers, and he soon found himself thrust into the hotbed of pre–Civil War secessionist politics. With relations between North and South steadily deteriorating, Lamar was elected to the United States Congress as a supporter of John C. Calhoun and Jefferson Davis. He had barely taken his seat in Congress before it became clear to all that Lucius Lamar was one of the most ardently pro-secession sons of the South then serving in Congress. He quickly gained a well-deserved reputation for being a firebrand who would brook no attempts by his Northern colleagues to cast aspersions on his beloved South. In fact, the most noted act of Lucius Lamar's pre–Civil War career came in 1861 when he drafted the ordinance of secession for the state of Mississippi. The pen of Lucius Lamar was the tool that severed Mississippi's bonds to the Union.

In the great conflagration begun at Fort Sumter, Lamar's family—like so many on both sides in the Civil War—suffered great losses. Of the 13 men descended from the first Lamar to set foot on American soil, 7 were killed in the war. Following the South's surrender at Appomattox, many of its now discredited leaders fled to other countries to escape what they feared would be the vengeful rule of their Northern conquerors, but Lamar chose to stay in Mississippi and brave the postwar privations to help rebuild his state. Lucius Lamar, who had led his state into war, now felt honor bound to help lead it to recovery. Like his fellow Mississippians, Lamar had to endure a Reconstruction period that was harsh and at times even degrading. For example, taxes during Reconstruction were 14 times what they had been before the war—and the taxes often just fattened the coffers of unsavory officials rather than improved the lives of those who paid them.

Abraham Lincoln had advocated a charitable rebuilding of the war-torn South and a coming together, without malice, of former adversaries, but his desire for reconciliation died with him at the hand of an assassin. Only Lincoln had the stature to rein in those in the North who wished to punish the South for secession. When he died, a vengeful Reconstruction was ensured. Some Southerners who did not hate the North before the Civil War came to hate it afterward as a result of the excesses of unethical occupation officials. These embittered Southerners often brought their complaints to Lamar, someone they felt sure shared their bitterness. They looked to him for redress of their grievances and relief from their suffering, but as a private citizen there was little Lamar could do. Consequently, in 1872, with the memories of Reconstruction fresh in their minds, Lamar's constituents returned him to the United States Congress.

Lamar took his seat in Congress determined to help his Mississippi constituents rebuild their lives, but his outlook concerning how best to accomplish this goal had begun to change. Lucius Lamar—once the most fire-breathing anti-Union secessionist—had come to believe that the South's only hope for restoration lay not in hate, but in reconciliation. "Lamar hoped to make the North realize that the abrogation of the Constitutional guarantees of the people of the South must inevitably affect the liberties of the people of the North. He came to believe that the future happiness of the country could only lie in a spirit of mutual conciliation and cooperation between the people of all sections and all states."[10]

Lamar decided that if a person like him, who had suffered so much for his anti-Union sentiments, could bring himself to extend an olive branch to his counterparts in the North, maybe other Southerners would be willing to follow his leadership, set aside their hard feelings, and work for reconciliation. Lamar knew he was risking the wrath of those who would not or could not put aside their deep-seated and bitter memories, but he also knew that if he did not step forward and set an example of conciliation, the South would continue to suffer for many years to come. He decided to do what no person—from North or South—could even imagine Lucius Lamar doing. He would risk the anger and disappointment of his Southern constituents by extending an olive branch to the North and accept the consequences, whatever they might be. Having made his decision, Lamar had only to find an appropriate occasion and venue. He was struggling with these issues when an answer unexpectedly presented itself—one of the

South's most vocal critics in Congress, Senator Charles Sumner, died suddenly. Seeing in Sumner's death the opportunity he had been searching for, Lamar decided to eulogize his former enemy on the floor of the House of Representatives.

To put into perspective the moral courage it took for Lamar to accept the invitation to eulogize Charles Sumner, consider the following background. Few people were more hated in the South than Charles Sumner of Massachusetts. Just as Lamar epitomized the anti-Union passions of secessionists, Charles Sumner epitomized the anti-Southern hostility of the most ardent abolitionists. Sumner was the radical Republican the South blamed more than anyone else for the harsh depredations of the Reconstruction period. This is the same Charles Sumner who labeled his Northern colleague, Daniel Webster, a traitor for trying to hold the North and South together. This is the same Charles Sumner whose anti-Southern rhetoric was so strident that Congressman Preston Brooks of South Carolina, incensed beyond restraint, actually beat him with a cane on the floor of the U.S. Senate—an act that made Brooks an instant hero throughout the South.

Understanding fully the personal risk he was taking, Lucius Lamar stood on the floor of the U.S. House of Representatives and delivered a eulogy for Charles Sumner so impassioned in its message of forgiveness and reconciliation that it brought tears to the eyes of even the most battle-scarred, jaded, and skeptical members of Congress. When Lamar had finished, his audience sat for several moments in shocked silence as if riveted to their chairs, unable to believe that the eloquently delivered words they had just heard had been spoken by Lucius Lamar. Then, almost as if by signal, the House chamber erupted in a loud and sustained ovation that sent shock waves reverberating out of the Capitol and throughout the country. Responses, both positive and negative and from both North and South, were almost immediate.

In Boston, Lamar was hailed as a "prophet of a new day in the relations between North and South."[11] Others in the North called the eulogy the most important and hopeful words to come out of the South since the end of the war. However, the response from Lamar's constituents back home was less encouraging. Several leading newspapers in his state claimed he had sold out to the North, surrendering his region's dignity, principle, and honor. The criticism hurt him deeply. Lamar loved his state of Mississippi and had sacrificed much in serving it. But as an effective leader, he felt compelled by conscience to take the long view and do what was right rather than what was popular. He would do so again and again in his illustrious but controversial career.

Lamar's own words, penned during some of his most difficult moments, speak not just for him, but for leaders in all times and from all fields who must choose between what is popular and what is right. He said, "I get many complimentary letters from the North, very few from Mississippi. . . .Can it be true that the South will condemn the disinterested love of those who, perceiving her real interests, offer their unarmored breasts as barriers against the invasion of error? . . . It is indeed a heavy cross to lay upon the heart of a public man to have to take a stand which causes the love and confidence of his constituents to flow away from him."[12]

Clara Barton[13]

One of the best examples of selfless leadership is Clara Barton. Barton is best known for her leadership in founding the American Red Cross, an organization that has, since its inception, come to the aid of hundreds of thousands of victims of war, famine, natural disasters, and other catastrophic events. However, had it not been for Barton's quiet but consistent example of selflessness over many years, there might not have been an American Red Cross. The selflessness lesson taught by the life of Clara Barton is as follows:

> **Selflessness can be a leader's most valuable asset in overcoming the resistance of people who stand in the way of the leader's vision.**

After several frustrating and underpaid years as a schoolteacher and bureaucrat, Barton found what turned out to be her calling when she helped nurse a nephew stricken with tuberculosis. Around this time—the spring of 1861—America's bonds of union were shattered by the outbreak of the Civil War. At the first Battle of Bull Run in July 1861, it became obvious to all concerned that the Union was not prepared for the unprecedented number of casualties that would be inflicted on the killing fields of this great war. The high volume of wounded quickly overwhelmed the Northern army's meager medical infrastructure.

Observing this unfortunate situation, Barton decided that something must be done, and that she would do it. She began by organizing charitable drives to supply Union troops with badly needed provisions and medical supplies, but soon found she wanted to do more. It was at this point that the selfless leadership for which Barton is known began to emerge. Knowing Barton felt a heavy burden to do more for the wounded, sick, and dying soldiers of the Union army, a friend suggested she go to the battlefield and nurse wounded soldiers there, something unheard of for a woman at the time. Although she met considerable resistance from skeptical army, government, and medical officials, Barton persisted and soon came to be known as "the angel of the battlefield" to the grateful troops under her care.

To care for the thousands of wounded on the battlefields of the Civil War, Clara Barton had to selflessly endure incredible hardships and privation, which she did with determination and quiet dignity. Her popularity with the troops and her propensity for going it alone while always questioning, prodding, and irritating bureaucratic officials to do more and do it better, brought Barton into frequent conflict with Union officials.

Although she was a paragon of idealism, Barton was realistic enough to recognize that her maverick ways were generating so much conflict and resentment among important officials that her effectiveness in caring for the wounded was being diminished. "Rather than persist in her maverick ways and risk being barred from the battlefield, Barton selflessly compromised and cooperated—at least to some degree—with officially sanctioned organizations."[14]

When the war ended, Barton organized searches for missing soldiers and led a project to identify and bury thousands of unnamed Union soldiers who had died in the Confederate prison camp at Andersonville, Georgia. She was also responsible for persuading government officials to convert the burial ground at this infamous prison into a national cemetery.

It was while giving lectures in Europe on her experiences in the Civil War that the International Convention of Geneva—more popularly known as the Red Cross—first came to Barton's attention. While she toured Europe, the Franco-Prussian War broke out (1870), and Barton volunteered her services to the Red Cross. Following this service, she worked at the German Red Cross Hospital in Baden. When she finally returned home to the United States in 1877, Barton began the work that would eventually lead to establishment of the American Red Cross.

"Barton embarked on a campaign to educate the American public on the international Red Cross movement, and she led a growing campaign to win acceptance for an American Branch of the organization. It was formed, at long last, on May 21, 1881."[15] Barton served as the first national president of the American Red Cross. Throughout her remarkable career, it was her selflessness—more than any other factor—that made Clara Barton such a persuasive, credible, and effective leader.

Ulysses S. Grant

The two best-known generals in the American Civil War were Ulysses S. Grant for the North and Robert E. Lee for the South.[16] Lee and his troops were outnumbered and outgunned by those of Grant, but through outstanding leadership, innovative tactics, and daring risks, Lee was able to hold Grant's army in check for many months. Grant was frequently frustrated and sometimes embarrassed by his inability to deliver a final and fatal blow to Lee's Army of Northern Virginia. However, although he knew that to win the war he would have to decidedly defeat Lee, Grant still admired the courage and tenacity of his adversary. The integrity lesson taught by one particular incident in the life of Grant is as follows:

> Leaders sometimes have to put their careers on the line to protect their integrity.

When Robert E. Lee finally surrendered his exhausted and starving troops at Appomattox, Virginia, Grant agreed to terms that were both honorable and magnanimous. There was no vindictiveness or punishment written into the surrender document. Grant knew that the Confederates still had an army in the field under General Joe Johnston that could conduct a guerrilla war out of the mountains of North Carolina for years to come. Consequently, he wanted a surrender document that would be the first step in the healing process—not a guarantee of a prolonged war. Predictably though, not all in the North agreed with Grant's views, and some of those who disagreed were highly placed, influential officials.

On June 7, 1865, a federal judge indicted Robert E. Lee for treason. Knowing that the officers and men of the South were protected by the United States government as long as they adhered to the terms of the surrender at Appomattox, and valuing the sanctity of the promises he made in the surrender document, Ulysses S. Grant was incensed at the charge brought against his former adversary.

Contacting the federal judge who had leveled the charge of treason, Grant insisted that the legal proceeding against his former adversary be stopped and that the conditions of the surrender document be honored. To ensure that the judge understood that he meant business, Grant actually threatened to resign his commission as general-in-chief of the United States Army if the charge was not dismissed. Then he forwarded a request to the president of the United States that Lee be pardoned. The charge of treason was dropped and the pardon was granted. Ulysses S. Grant's integrity stopped an ill-advised move by a federal judge that would have rekindled the still-burning embers of sectional strife and delayed even further the healing of America's wounds.

John Quincy Adams[17]

John Quincy Adams is typically remembered as the sixth president of the United States and as the younger half of the only father and son team to serve as presidents prior to George Herbert Walker Bush and his son George W. Bush. But John Quincy Adams should also be remembered as a leader who served his country long and well with unflagging integrity. Although it occurred so long ago—the late 1700s and early 1800s—the service of Adams to the United States has never been equaled.

When he died at 80 years of age, John Quincy Adams had held more public offices than any other American. His record still stands. Those offices include state senator from Massachusetts, minister to the Hague, emissary to England, minister to Prussia, minister to Russia, head of the American Mission to negotiate peace with England (War of 1812), secretary of state, member of the United States House of Representatives, United States senator, and president of the United States. An interesting historical sidebar is that Adams is also the only American president to be elected to Congress *after* serving as president.

What makes the career of John Quincy Adams so interesting is not just that he held so many offices but that he was able to do so in spite of frequently putting himself at odds with his party and his constituents in order to preserve his integrity. The primary lessons taught by the career of Adams are as follows:

> What is right in a given situation is not changed by the personal consequences a leader might suffer for doing it.
>
> A leader with integrity must put being respected ahead of being liked.

John Quincy Adams was a man of integrity and principle who believed that his country came before his party and that right came before expediency. These beliefs cost him dearly. In the first 50 years of America's existence as a country, only two presidents failed to win election for a second term—John Quincy Adams and his father. In a letter to his father, Adams once wrote, "I have already had occasion to experience, which I had before the fullest reason to expect, the danger of adhering to my own principles. The country is so totally given up to the spirit of party that not to follow blindfolded is an expiable offense."[18] Adams knew the price he would pay for adhering to his principles because he had seen his father pay the same price for adhering to the same principles.

What so often made John Quincy Adams a "minority of one" was his commitment to put the good of his country ahead of the more parochial interests of his party, state, or region. Examples of his courageously stubborn independence abound. Soon after being elected to the Massachusetts state legislature as a member of the Federalist Party, he enraged party leaders by proposing that the Jeffersonian Party—archenemies of the Federalists—be given proportionate membership in the governor's cabinet. Later, while serving in the United States Senate, Adams brought the enmity and scorn of his Federalist colleagues down squarely on his shoulders by supporting President Thomas Jefferson's proposal to purchase the Louisiana Territory from Napoleon's France.

The Federalists and most people in New England opposed the Louisiana Purchase out of fear that the westward expansion it represented would diminish the political and economic hegemony then enjoyed by the northeastern states. But Adams thought the purchase would be good for the United States because it would remove France from America's continental boundaries while helping America expand to reach those boundaries. The wrathful denunciations of his colleagues and constituents following this incident were so severe that few public servants could have endured them, but Adams, although hurt by the criticism and threats, persevered and did so without bending. In his diary, Adams wrote about those "who hate me rather more than they love any principle."[19] About one particularly contemptuous colleague, Adams wrote that he "abandons altogether the ground of right, and relies upon what is expedient."[20]

The most controversial and unpopular stand by Adams, and probably his most courageous, came during the years leading up to the War of 1812 with Great Britain. During this time, American cargo ships were frequently stopped on the high seas by British men-of-war only to have their cargoes confiscated and their crews impressed and compelled to serve in the British navy. Adams considered these despicable acts to be nothing short of piracy, but the Federalist merchants whose ships were being raided preferred to appease Great Britain rather than stand for the rights of the Americans who were being taken hostage. The Federalist appeasers were more afraid of losing British business than of losing the occasional crew and cargo. They preferred to simply write off their losses as business expenses and continue the status quo.

Such an expedient attitude brought nothing but contempt from Adams, who in 1806 introduced a series of resolutions condemning Great Britain for its aggression and demanding restoration of all ships, men, and goods lost. Then he went even further by supporting a bill limiting British imports. One of the least-inflammatory responses of the Federalists to this action was a public statement claiming, in part, that Adams should "have his head taken off for apostasy . . . and should no longer be considered as having any communion with the party."[21]

In 1807 President Thomas Jefferson believed he could no longer let British actions go unpunished. In response to their continued aggression on the high seas, Jefferson proposed an embargo on all trade with Great Britain, a measure that would have a devastating effect on the economy of Massachusetts. Not only did Adams, against the bitter denunciations of his Massachusetts constituents, support the president's bill, he introduced one of his own that went even further. Having done so, Adams remarked to a fellow senator, "This measure will cost you and me our seats, but private interest must not be put in opposition to public good,"[22] a courageous stand for a courageous leader who valued his integrity.

Bill Gates[23]

Most people in the world of business know the story of how Bill Gates, CEO of Microsoft and perennially the wealthiest man in America, dropped out of Harvard to found what would become one of the most successful software development firms in the world. But less is known about the selflessness of this dynamic leader of the personal computer revolution. The selflessness lesson taught by the example of Bill Gates is this:

> The best leaders are those who give back some of the fruits of their success.

Gates is one of the most philanthropic corporate executives in the world:

Philanthropy is also important to Gates. He and his wife, Melinda, have endowed a foundation with more than $24 billion to support philanthropic initiatives in the areas of global health and learning, with the hope that as we move into the 21st century, advances in these critical areas will be available for all people. To date, the Bill and Melinda Gates Foundation has committed more than $2.5 billion to organizations working in global health; more than $1.4 billion to improve learning opportunities, including the Gates Library Initiative to bring computers, Internet [a]ccess and training to public libraries in low-income communities in the United States and Canada; more than $260 million to community projects in the Pacific Northwest; and more than $381 million to special projects and annual giving campaigns.[24]

Gates comes by his philanthropy naturally. His late mother, a schoolteacher, once served as chair of United Way International. It would be easy for the world to resent Bill Gates for his incredible wealth, but it is difficult to resent a leader who gives so much and so selflessly to so many different needy and worthy causes. Because of his giving, Gates has never been portrayed negatively by the media in the same way that some business leaders of wealth are on occasion. Showing both philanthropy and wise leadership, he realizes that those with more wealth also have more responsibility to be good stewards of it.

Summary

1. Integrity is strict adherence to a code of ethics that a person has internalized. A person of integrity has adopted a positive value system, knows what beliefs make up that system, and lives according to those beliefs regardless of circumstances, peer pressure, present company, or other influences.

2. People with integrity have in common the following traits: refusal to pretend, internal moral compass, consistency in behavior, consistency in decisions, concern for substance over image, and selflessness.

3. A selfless person puts the needs of the organization and the people he leads above his own. People with integrity are invariably selfless, because acts of integrity are by their very nature selfless. When an individual risks his personal interests to do the right thing, as opposed to the popular thing, he is performing a selfless act.

4. Integrity is important because it builds trust, gives the leader influence, establishes high standards, establishes a solid foundation rather than just an image, and builds credibility.

5. Ethical behavior means doing the right thing as *right* is defined within a given value system. Leaders are continually faced with ethical dilemmas in which they must balance the needs of individuals and the needs of the organization. Ethical dilemmas often put leaders in a kind of double jeopardy in which there are problems regardless of which choice

they make. Leaders who are trying to make a decision that has high ethical content can use the following tests: morning-after, front-page, mirror, role-reversal, and commonsense tests.

6. No human being is likely to ever achieve perfect integrity. This should be understood by those who would lead, but it should never be used as an excuse to stop striving for perfection. The closer a leader comes to perfect integrity, the better he will be able to lead others.

Commonsense test	Integrity establishes a solid foundation
Concerned more with substance than image	Integrity establishes high standards
Consistency in behavior	Integrity leads to influence
Consistency in decisions	Internal moral compass
Double jeopardy	Mirror test
Ethical behavior	Morning-after test
Front-page test	Refusal to pretend
Integrity	Role-reversal test
Integrity builds credibility	Selflessness
Integrity builds trust	

Review Questions

1. Define the concept of integrity as it relates to leadership.
2. List and explain the characteristics people with integrity share.
3. Explain what is meant by selflessness and give an example of the concept.
4. List and explain the reasons that integrity is so important to those who would lead.
5. Explain the double jeopardy leaders sometimes face when making decisions that have high ethical content.
6. List and explain the tests leaders can use to decide the most ethical course of action.
7. What is the most difficult factor to overcome when trying to make an ethical decision?

LEADERSHIP SIMULATION CASES

The following simulations are provided to generate additional thought and discussion about the principles of leadership explained in this chapter. Readers are encouraged to consider how the situations presented in these cases might apply to them and to discuss the cases with other leaders and leadership candidates.

CASE 2.1 | I Don't Trust Our New CEO

"What do you think of our new CEO's vision?" asked Margaret Colby. She and her colleague Vince Gardner were having a cup of coffee after attending a companywide meeting in which Aero-Tech, Inc.'s new CEO laid out his vision for their company. The two engineers had worked at Aero-Tech for 10 years under one CEO, the company's founder, who was admired, respected, and trusted by all 410 of the company's employees.

"His vision sounded fine," responded Gardner. "But I don't trust him." "Why not?" asked Colby, obviously surprised by her colleague's answer. Gardner explained that he played tennis with an engineer who used to work for their new CEO at another company. According to this colleague, the new CEO of Aero-Tech was a "climber" who just used companies as stepping-stones on his way to bigger and better things. "There is nothing wrong with ambition," said Colby. "No, there isn't," agreed Gardner. "But there is something wrong with making self-interested decisions that work out well in the short run but actually hurt the company in the long run—after the CEO has moved on to his next company, or should I say 'victim'?"

Discussion Questions

1. Have you ever worked in an organization led by someone you did not trust? If so, what effect did your lack of trust have on your morale and productivity?

2. What kinds of problems might the new CEO of Aero-Tech have in realizing his vision for the company if he fails to gain the trust of his employees?

CASE 2.2 An Ethical Dilemma

Waste Removal, Inc. (WRI), has several teams of drivers who transport compacted solid waste to the company's landfill. WRI is a responsible company that has invested a great deal of time, expertise, and money in developing a new, environmentally sound landfill that can safely handle all of the various types of solid waste the company accepts. There is, however, one major problem with the landfill—its distance from the company's compacting plant.

Drivers complain that the 100-mile round trip for each load limits their ability to earn the financial incentives for exceeding their weekly tonnage quotas that they could earn at the old, nearer landfill. These financial incentives are at the heart of Mack Ritter's ethical dilemma. Ritter leads WRI's collection and transporting department. All of the drivers who collect solid waste and those who transport it to the landfill report to Ritter.

One of Ritter's best, most loyal, most experienced drivers desperately needs to earn the financial incentives available to him. This driver, Jason Caplan, has a child in the hospital. Even with insurance, the child's treatment is stretching Caplan's personal finances to the breaking point. Ritter has learned that to maximize his incentive pay, Caplan is dumping every other load he transports in WRI's old landfill. This landfill, which is still open but only for compost-type materials, is not rated to handle some of the potentially hazardous solid waste WRI accepts. Ritter is torn between the needs of a loyal employee and the potential harm to his company in lawsuits, not to mention the potential damage to the environment of his community.

Discussion Questions

1. What do you think is the right thing for Ritter to do in this situation?
2. What problems might Ritter face if he confronts Caplan with what he knows? What problems might Ritter face if he just ignores what Caplan is doing and lets it continue?

CASE 2.3 The Selfish Manager

"I cannot believe these workers' compensation costs," shouted Angela Petry. "This is the third quarter in a row they have increased. It seems like half of the employees in my department are on some form of light or restricted duty." Since rising up through the ranks to become manager of the information technology department of Whalen Manufacturing Company (WMC), Petry had experienced numerous instances of reluctant compliance and tactical disobedience by her direct reports. As a result, she suspected that her ever-increasing workers' compensation costs were being caused by perfectly healthy but "whiny" employees pretending to have minor injuries such as tendonitis and carpal tunnel syndrome.

When she confided her suspicions to a friend who worked in another department at WMC, Petry was shocked at what the friend told her. She said, "You're probably right in your suspicions. I hear things from time to time, and if only half of what I hear is true, I don't blame your employees for lying down on the job." Her friend went on to tell her that over the years she had developed a reputation for caring about no one but herself. They knew she cared about getting the company's work done, but they did not think she cared about them. Her direct reports saw her as a person who would willingly ask them to work weekends but would never come in herself and help them. They saw her as a person who would gladly step on them as she climbed her personal career ladder.

Discussion Questions

1. Do you think Petry will ever be able to effectively lead the information technology department without changing the image her employees have of her?

2. How might Petry go about changing her negative image with the employees in her department?

Endnotes

1 *God's Little Devotional Book for Leaders* (Tulsa, OK: Honor Books, 1997), 303.

2 God's *Little Devotional Book*, 303.

3 John C. Maxwell, *Developing the Leader within You* (Nashville, TN: Thomas Nelson, 1993), 38–44.

4 James M. Kouzes and Barry Z. Posner, *The Leadership Challenge*, 3rd ed. (San Francisco: Jossey-Bass, 2002), 24–27.

5 Louis E. Boone, *Quotable Business*, 2nd ed. (New York: Random House, 1999), 31.

6 Thomas J. Neff and James M. Citrin, *Lessons from The Top* (New York: Currency & Doubleday, 1999), 327–332.

7 Louis E. Boone, 194.

8 John F. Kennedy, *Profiles in Courage* (New York: Harper & Row, 1956), 159–185.

9 John F. Kennedy, 185.

10 John F. Kennedy, 169.

11 John F. Kennedy, 161.

12 John F. Kennedy, 181–182.

13 Alan Axelrod, *Profiles in Leadership* (Upper Saddle River, NJ: Prentice Hall, 2003), 61–64.

14 Alan Axelrod, 62–63, 185.

15 Alan Axelrod, 63–64.

16 Al Kaltman, *The Genius of Robert E. Lee* (Upper Saddle River, NJ: Prentice Hall, 2000), 322.

17 John F. Kennedy, 37.

18 John F. Kennedy, 43.

19 John F. Kennedy, 46.

20 John F. Kennedy, 46.

21 John F. Kennedy, 48.

22 John F. Kennedy, 48.

23 www.microsoft.com/billgates/bio.asp

24 www.microsoft.com/billgates/bio.asp

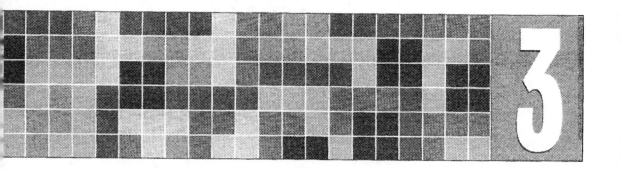

Establish Credibility and Good Stewardship

"All organizations are at least 50 percent waste—waste people, waste effort, waste space, and waste time."[1]

—Robert Townsend,
Former President, Avis Rent-a-Car

OBJECTIVES

- Define credibility in the context of leadership.
- Explain why credibility is so important to those who would lead.
- Explain how to establish credibility with those you want to lead.
- Explain how to maintain credibility once you have established it.
- Define the concept of stewardship in the context of credibility.
- Summarize the credibility and stewardship lessons from selected effective leaders.

So much of leadership is tied to the concept of followership, which is a willingness to follow another person's lead in pursuing a vision, goals, or plan. A leader without followers is no leader at all, and a leader without credibility will have no followers. According to James Kouzes and Barry Posner, "*Credibility is the foundation of leadership. Above all else, we must be able to believe in our leaders. We must believe that their word can be trusted, that they'll do what they say, that they are personally excited and enthusiastic about the direction in which we're headed, and that they have the knowledge and skills to lead.*"[2]

CREDIBILITY DEFINED

Credibility is "the quality, capability, or power to elicit belief."[3] To have credibility, people who hope to lead must have several critical characteristics:[4]

- Honesty
- Ability to be forward-looking
- Competence
- Ability to inspire

WHY IS CREDIBILITY SO IMPORTANT TO LEADERS?

To succeed in today's intensely competitive global marketplace, organizations need consistent peak performance from their personnel. To perform at peak level consistently, employees must be as committed as management to the organization's goals. This is why credibility is so important to leaders. Employees are more likely to make a willing commitment to follow a leader they believe in—one who has credibility. They will reluctantly go along with the directions of a manager who lacks credibility, but they will not put their heart into the work. The difference between employees who willingly commit to a leader and those who just reluctantly comply can be the difference between winning and losing in a competitive marketplace.

According to Kouzes and Posner,[5] employees who believe in their immediate manager or supervisor are much more likely to

- Take pride in their organization.
- Exhibit team spirit.
- Believe that their values match those of the organization.
- Feel they are a part of the organization.
- Make a commitment to the organization.

Employees who do not believe in their immediate manager or supervisor are more likely to:[6]

> ### Leadership Tip
>
> *"It isn't the incompetent who destroy an organization. The incompetent never get in a position to destroy it. It is those who have achieved something and want to rest upon their achievements who are forever clogging things up."[7]*
>
> —Charles Sorenson
> American Manufacturing Executive

- Work hard only when they are observed.
- Derive their motivation only from money.
- Criticize the organization when in private.
- Pursue other employment quickly when the organization experiences problems.
- Feel they are unappreciated and unsupported.

Clearly, credibility has a significant effect on an organization's ability to compete. Productivity, quality, customer satisfaction, and the continual improvement of these critical factors can be linked directly to the credibility of those who lead.

HOW TO ESTABLISH CREDIBILITY

As stated earlier, the credibility of a leader depends to a large extent on four key factors: the leader's honesty, ability to be forward-looking, competence, and ability to inspire followers. John Maxwell expands these four bottom-line characteristics into seven as follows.[8]

Character

To have credibility, leaders must be perceived by stakeholders as being honest, ethical, and trustworthy. People know they can believe in leaders who have these characteristics. Conversely, they are reluctant to follow a leader who lacks honesty, ethics, and trustworthiness, because they cannot trust where he might lead them. People will not willingly follow someone they do not believe in.

John X had amassed great personal wealth by becoming a turnaround expert in the field of manufacturing. He specialized in taking over struggling manufacturing firms and turning them into winners. His formula was simple—cut costs to the bare bone and, by so doing, enhance the bottom line while undercutting the competition. His formula worked in the short run, and a short period is as long as he stayed with a given company. As soon as the company appeared to be well on the way to recovery, John

X would take his incentive money and move on to another struggling firm.

Eventually, after John X had "saved" four companies, sufficient time had elapsed to allow stakeholders to measure the long-term effects his turn-around strategies were having. His long-term record finally caught up with him and led to John's downfall. What had appeared to turn the companies around in the short run had actually done them more harm than good in the long run. John was almost indiscriminate in cutting costs and termi-nating employees—employees he had promised to retain if they would stay on and help him turn the company around.

Unfortunately for the long-term good of the companies, John's cost-cutting measures eliminated employees and technologies that were essen-tial to the long-term competitiveness of the companies. As a turnaround expert, John X was like the farmer who raises some quick cash by selling off his seed corn. As a result, when spring comes he is unable to plant a new crop.

Word of the long-term effect of John's so-called turnaround methods and his dishonesty with employees eventually began to spread. As a result, when he took over his fifth company, employees at all levels were reluctant to follow his lead. In fact, many key employees who John needed in order to work his "magic" quickly left to join competitors. In spite of his rhetoric to the contrary, John's track record showed he could not be trusted. His fifth turnaround was a disaster—for him and for the company. John X, once the brightest star in the world of manufacturing, quickly faded because he had no credibility with those he hoped to lead.

Relationships

Those who want to lead other people must get to know those people, and they must become known by them. Much of a leader's success is tied to the relationships he has with his followers. People will follow a person more readily than an idea or a vision. Following someone is a deeply personal decision. We tend to believe in people first and in ideas second. It is much easier for us to accept the vision of someone we believe in than someone we don't, regardless of the apparent validity of his vision. Consequently, those who hope to lead must invest the time and effort necessary to es-tablish positive, trusting, mutually supportive relationships with potential followers.

Knowledge

Followers look to leaders for guidance and for answers. Nobody wants to fol-low someone who appears to be ill-informed, unprepared, and lacking in-dustry knowledge. This is one of the reasons the author rejects the idea that a good manager can manage any kind of organization. Although this might be partially true, the fact is that if two leaders have equal management ability, but one of them is knowledgeable in the industry and one is not, the more knowledgeable leader will be the better leader because he will have credibility.

Intuition

The ability to deal with the all-pervasive intangibles of working in a competitive environment is a must for leaders. When leading people in circumstances that can change by the minute, there are more exceptions to the rules of business than there are rules. Ambiguity is an ever-present fact of life for those who lead organizations that compete in the global marketplace, and being able to deal effectively with ambiguity is essential to success. After all, if the target were stationary, anybody could hit it. The ability to hit a moving target in terms of what strategies and what decisions make the best sense is one of the key factors that separates winners from losers in the global marketplace.

Making sense out of the many intangibles faced by today's business leaders requires intuition—perceptive insight unaided by definite facts. People with intuition have a feel for what represents the best decision in ambiguous circumstances. Correspondingly, employees have a feel for which leaders have intuition, and they are more likely to follow those who do.

Experience

Few things can take the place of experience in helping leaders establish credibility. Followers like to know that a leader has "been there and done that." Engineers with solid experience in their field will typically find it easier to establish credibility with other engineers in that field. The same can be said for the experience of any technical professional. The closer the leader's experience comes to that of potential followers, the better, because common experience gives leaders and followers common ground, a major plus when trying to establish credibility.

Past Success

People like to follow a winner, and their rationale for this preference is simple: If he has won before, he can win again. This principle shows up every year in the fruit basket turnover of college and professional coaches. Those who have won consistently are offered better and bigger positions. Those who are seen as losers are out and their prospects are dim. A good track record is critical to those who would lead.

Ability

A leader who can do what he expects his followers to do, and do it well, will find it much easier to establish credibility than someone with no pertinent ability. In addition to field-specific ability, those who would lead must also be good at their current jobs. For example, if a leader's position is vice-president for engineering, he should be a good engineer, but he should also be good at the management tasks that go with the job (organizing, planning, leading, controlling, mentoring, etc.).

Followers are very sensitive to the abilities of those who want to lead them. A leader will have difficulty inspiring followers to make a willing, total commitment to the organization's goals if he is viewed as being incompetent, or even just marginally competent. In addition, the leader must win followers' respect. Employees will find it difficult, if not impossible, to respect someone whose competence they question.

**Five Rules for
Maintaining Credibility**

1. Set the example (practice what you preach).
2. Support your people (take the blame and share the credit).
3. Admit mistakes (learn from mistakes and move on).
4. Follow through (do what you say you will do).
5. Be consistent (be equitable and fair with all employees).

FIGURE 3.1   Credibility must be maintained or it can be quickly lost.

HOW TO MAINTAIN CREDIBILITY

Credibility is like a suntan—it will quickly fade if not maintained. Figure 3.1 lists the author's five rules for maintaining credibility.

Set the Example

To maintain credibility with their followers, leaders must do more than just "talk the talk"; they must also "walk the talk." The most credible leaders are those who consistently exemplify the principles they advocate. If you want followers to be on time, you must be on time yourself. If you want followers to adhere to a corporate dress code, you must be a walking model of that code. If you want followers to treat customers in a certain way, you must treat customers—as well as the followers—in this way.

Leadership Profile   **Bob Eaton and the Daimler-Chrysler Merger**[9]

The merger of Daimler-Benz—maker of Mercedes-Benz—and Chrysler—maker of Dodge, Plymouth, and Jeep—took many in the business world by surprise. After all, Daimler-Benz was an old-school European producer of upscale vehicles, whereas Chrysler was a perennially struggling producer of mass-market vehicles. Theirs was viewed as a strange marriage by many, and some business analysts wondered aloud if the merger was ill-advised. But the doubters, skeptics, and naysayers did not reckon with Bob Eaton, who at the time of the merger was chairman and CEO of Chrysler.

Daimler-Benz officials saw the merger with Chrysler as a way to reach a younger market while simultaneously expanding imports. Eaton saw the merger as a way to achieve his vision of transforming Chrysler from a company perennially struggling to stave off bankruptcy into the premier car maker in the world. But for the merger to work, it would be essential for Chrysler's most critical stakeholders (management personnel, technical professionals, union workers, dealers, and customers) to follow Eaton's lead; and follow they did.

In fact, Eaton regards leadership as the ability to take stakeholders somewhere they need to go but don't necessarily want to go, and this is precisely what he did with the Daimler-Chrysler merger. Chrysler's key stakeholders did not know if the merger was a good idea, but they did know that Bob Eaton thought it was; and that was good enough for them because they believed in Eaton.

How Eaton earned this level of credibility with Chrysler's stakeholders is instructive for technical professionals who want to be leaders in their organizations. First, Eaton was well known to stakeholders as a car man. He had spent almost 20 years in the automobile business before becoming Chrysler's CEO. Talking about cars and driving them were equal passions for this car-loving executive. Eaton was like the owner of fine racehorses who, before becoming an owner, had both trained and ridden them. Eaton's competence in automotive engineering, product development, and process improvement were already legendary when he took over the reins at Chrysler. Eaton knew his business, and Chrysler's stakeholders knew he knew.

In addition to competence, Eaton had a well-earned reputation for being forward-looking. Before the merger with Daimler, Eaton had taken steps to refine Chrysler's product development process, increase its exports, and attract a younger market. All of these things had made Chrysler much more competitive in the global marketplace, which is why Daimler was interested in the merger in the first place. His competence and ability to look forward for the benefit of the company gave Bob Eaton the credibility he needed to lead Chrysler through a successful merger with Daimler.

Where some leaders go awry is in thinking that because they have risen to a position of responsibility and authority, the rules no longer apply to them. This exalted attitude is as contrary to real leadership as it can possibly be. Rather than releasing them from obligations, leadership imposes even greater obligations on those in positions of responsibility and authority. One of those is the unrelenting obligation to set a positive example.

Support Your People

An unwritten rule that applies equally in the worlds of commerce, politics, sports, and the military is that the leader gets the blame when things go wrong and the credit when things go right, even if he does not deserve it in either case. Leaders need to be aware of this unwritten rule and understand that it applies whether they like it or not. This might seem unfair and probably is, but so is life; and life is often doubly unfair in the case of leaders. However, if you understand this unwritten rule, you can use it to maintain credibility. Here's how. When things do not work out the way you want them to, even if the fault clearly lies with members of your team, as the leader you should step forward and take the blame. After all, you are going to be blamed anyway, so you might as well at least add to your credibility by deflecting the blame from your team members.

The obverse is also true and equally important: When things work out well, the leader is going to receive the credit. At this point, it is critical that you share the credit with your team members. Those who deserve the credit will appreciate it, and those who do not will work even harder in the future to make it up to you. When credit is being given, leaders should always say "we," avoiding "I" completely.

Admit Mistakes

Mistakes are going to occur. Leaders make decisions based on the best information they can gather, on whatever intuition they may have, and within the time frame imposed by circumstances. Unfortunately, information is sometimes inaccurate and, due to time constraints, often incomplete. Intuition is helpful, but people are not omniscient. And time almost always works against you in the decision-making process. It is like money—when you need it the most, you often have it the least. Consequently, leaders are going to make decisions that do not work out.

This fact does not excuse leaders who make poor decisions, nor should it be used as an excuse for problems that arise from faulty preparation or insufficient planning. Leaders are responsible for making the best decisions they can make given the circumstances of the situation in question.

However, when you have thoroughly prepared, properly planned, and considered all of the information that can be gathered within the time available, and your decision still does not produce the desired result, the best approach is to simply admit the mistake openly and honestly. Then, having done so, learn from your error and do better next time.

Everybody makes mistakes, even the best of leaders. Those who openly admit their errors will be more credible in the eyes of stakeholders than those who try to point the finger of blame elsewhere or who make whiny excuses. In fact, the evidence on this topic suggests that making an occasional error, readily admitting the mistake, and taking full responsibility for it can actually enhance a leader's credibility. This does not mean that leaders should err on purpose just so they can stand tall before their followers and admit their mistake. On the contrary, leaders will make plenty of mistakes in the normal course of events without needing to artificially induce them. It is not the making of mistakes per se that adds to the leader's credibility; rather, it is how leaders handle their mistakes that will cause them to win or lose followers.

An example from history illustrates the truth of this principle. In the months before John F. Kennedy was elected the 35th president of the United States, his predecessor, Dwight D. Eisenhower, had approved a plan by the Central Intelligence Agency to support an invasion by Cuban expatriate freedom fighters of their home island, now controlled by the communist dictator Fidel Castro. A key element in the CIA's support of the invasion was air strikes to be provided covertly by the American military. Kennedy knew little about the planned invasion when he took office. But shortly thereafter, the CIA gave him a briefing. By now the invasion was imminent. Consequently, as president, Kennedy had to be given the details.

The invasion came at a time when relations between the United States and the former Soviet Union—Castro's principal ally and supporter—were on difficult footing and getting worse almost by the day. The threat of a nuclear war with the Soviet Union was a clear and present danger. Kennedy was concerned that American support of a Cuban invasion might trigger a Soviet move on West Berlin, but the CIA pulled out all stops to convince the new president that the invasion had to be allowed to proceed. Kennedy compromised. He approved the invasion but allowed only the first air strike. He called off a second strike, without which the invaders had little chance of success.

The Cuban freedom fighters who rushed ashore at the Bay of Pigs to wrest their country from the hands of a communist dictator were either killed on the beach or quickly captured and hurried off to endure the vengeful realities of Castro's prisons. With air support from the United States, the invasion might have succeeded; but without air support, it was doomed to failure from the outset.

In the aftermath of the failed Bay of Pigs invasion, the Soviet Union and its Cuban puppet, Castro, went out of their way to use the tragedy to embarrass President Kennedy in an attempt to undermine his presidency. Kennedy's handling of the situation is a textbook example of how leaders should confront their mistakes. Kennedy secured time on national television and forthrightly took full responsibility for the debacle. He could have said that President Eisenhower and the CIA dumped the plans for the invasion in his lap as a fait accompli. He could have said that he was not given sufficient time to study the plans and make a better decision. He could have said these things, and he would have been right. Instead, he looked directly into the camera and took full responsibility for a terrible mistake.

What those who would lead should know about this historical event is that following his televised admission of a major error in judgment, President Kennedy experienced one of the highest public approval ratings of his entire presidency. The American people showed they can relate to someone who makes mistakes—they make them too. But even more important for technical professionals who want to be leaders, they showed that they admire someone who has the moral courage to admit his mistakes and take responsibility for them.

Follow Through

Following through means doing what you say you are going to do. It means keeping promises. Two rules of thumb to remember for maintaining credibility are (1) promise only what you can deliver, and (2) promise small but deliver big. Nothing will undermine your credibility faster than failing to keep promises. Followers want to know without the slightest doubt that they can depend on their leaders. Consequently, it is better to disappoint people up front when you cannot deliver what they want than to make a promise you cannot keep.

Along these lines, because followers are so acutely aware of the leader's performance in delivering on promises, it is better to promise small but deliver big. When people want something from you, there is a natural human tendency to respond to the pressure by promising more than you can deliver or promising to deliver it earlier than is actually practical. Consider, for example, the following case. The engineers in Maria Padrone's division had been pressing her hard to secure sufficient funding to upgrade their solid modeling software. So far, Padrone had been put off twice by her boss, who told her, "The budget is too tight right now—maybe later."

With her engineering staff becoming increasingly frustrated, Padrone was feeling pressure to get results. Consequently, when her boss finally told her they might be able to carve out some funding, Padrone made the mistake of promising her engineers she would deliver the software upgrade they needed within a month. As it turned out, what she was able to deliver was a partial upgrade, and it took three months. As a result, she lost credibility with her staff. It would have been better for Padrone to promise small. For example, until she knew for sure how much funding would be allotted her department—and when—she should have made no promises. Then when she knew for certain how much and when, she should have given her staff a smaller figure and a later delivery date. In this way she would have built in a contingency should the funds allotted be less than expected or the date later than expected—both of which are common occurrences in any organization. Further, if the amount of funds and the delivery date turned out as expected, she would have exceeded her promise, thereby gaining credibility instead of losing it.

To understand how important it is to promise small but deliver big, consider whether you trust the word of airline ticketing agents when a flight is delayed. When they reschedule the flight for 30 minutes later, do you really trust their word? Those who travel frequently by air know that some airlines are prone to make promises they do not and often cannot keep. As a result, many flyers do not trust what they are told by the airlines, particularly when it comes to take off and arrival times. Almost everybody who flies has had the experience of having a flight delayed, only to sit in an airport terminal for hours while the airline updates its promise to leave soon every 30 minutes, until, finally, the flight is canceled altogether—and then only after it is too late to catch another flight to the desired destination. Airlines who do this are guilty of promising big but delivering small, a guaranteed formula for losing credibility with travelers.

STEWARDSHIP AND CREDIBILITY

Effective leaders are good stewards of all of the resources for which they are responsible—human, technological, financial, and physical. According to Peter Block, "Stewardship is to hold something in trust for another. Historically, stewardship was a means to protect a kingdom while those rightfully in charge were away, or, more often, to govern for the sake of an underage king. The underage king for us is the next generation.... It is the

---

Leadership Tip

"A promise made is a debt unpaid." [10]

—Robert Service
Canadian Writer

willingness to be accountable for the well-being of the larger organization by operating in service, rather than in control, of those around us."[11]

Stewardship means leading in a way that ensures the organization will be in better condition when you leave than when you started. This is the "next generation" aspect of stewardship. Leaders who are good stewards take care of the resources entrusted to them and use them wisely, effectively, and efficiently.

Leaders who are good stewards add to their credibility because their followers know that the leaders are committed to taking care of them and the organization. Leaders who are good stewards lead through service to their followers. This service manifests itself in many ways. Good stewards roll up their sleeves and pitch in to help when the workload backs up. Good stewards play an active and positive role in the career development of their followers. They get to know their followers personally and help them strike an appropriate balance between work and family or other personal obligations. Good stewards do what is necessary to continually improve their people and their processes. They manage their budgets efficiently and use their technologies effectively.

Betty X is director of software development for Safety Software, Inc. (SSI). She is also a good leader and a good steward. It is the nature of her business that her software developers sometimes have to work long hours over an extended period of time to meet deadlines imposed by the marketplace. Whenever OSHA changes a safety standard, SSI must very quickly update its software and get the updated version out to all customers. Betty X makes sure that all employees in her department know about this aspect of the job before they are hired. She also gets to know them well enough to understand the family obligations and situations of each technical professional in her department.

Betty X sends letters to family members to let them know about the occasional long hours and to apologize for interruptions to family time. In addition, whenever long hours are required, Betty pitches in and works right beside her software developers. When work hours return to normal, Betty schedules paid compensatory time off from work and gives her employees various types of family-oriented perks such as movie tickets, gift certificates to restaurants, and expense-paid weekend getaways. She also monitors her employees carefully to identify those who might be experiencing family problems as a result of the long hours. To the extent possible, she and the other employees cover for each other when, during periods of long work hours, a child is in a school play, wins an award, or has an important recital or ball game.

Because she is a good steward, Betty X has a high level of credibility with her direct reports at SSI. They know she cares about them and the company and that she will make every effort to maintain an appropriate balance between the needs of both.

LESSONS ON CREDIBILITY FROM SELECTED LEADERS

Following are excerpts from the lives of several leaders that exemplify some of the principles of credibility set forth in this chapter. The leaders selected for inclusion are Henry Harley "Hap" Arnold, Sitting Bull, Francis Turner, and Margaret "Meg" Whitman.

Henry Harley "Hap" Arnold[12]

Henry Harley Arnold was known to his contemporaries simply as "Hap." Arnold is known to students of leadership as "the father of the United States Air Force." He was born in Gladwyne, Pennsylvania, in 1886, long before the Wright Brothers proved that manned flight was possible. As a graduate of West Point and an infantry officer, Arnold was steeped in the old-school army traditions. But he was a precocious thinker and innovator who saw in the earliest days of human flight its military possibilities. The career of Hap Arnold teaches the following lesson:

> With credibility and perseverance a leader can achieve a lofty vision.

This visionary leader made several historic contributions to creating what is now the most advanced technological military force in the world—the United States Air Force. First, Arnold saw the military potential of manned air flight. Having seen the vision, he became a persistent advocate of the concept, personally mastered the science of aviation, became a leader among aviation advocates, and used his knowledge and influence to "*father*" the United States Air Force.

Arnold began trying to convince the army of the potential of a separate, independent air arm while many in this tradition-bound branch of the military were still riding horses—a tough sell on a good day. To put into perspective how difficult his task was, consider that at the time he was trying to sell the army brass on airplanes, most did not yet even accept such ground-oriented innovations as the tank. However, Arnold was nothing if not determined.

Arnold knew he would have to establish unquestionable credibility with army leaders before he would have the slightest hope of persuading them to listen to his theories. He also knew that he would have to demonstrate the value of an air arm in ways that even the most jaded skeptics could understand. He began by transferring from the infantry to the aeronautical section of the Army Signal Corps. Then he volunteered for flight training with the famous Wright Brothers at Dayton, Ohio.

Having learned to fly, Arnold demonstrated to army decision makers how the airplane could be used for reconnaissance. This was an important step because reconnaissance—the process of determining where the enemy is, where he is going, what he is doing, and how strong he is—is critical to ground commanders. Arnold had found a way to show ground-oriented army officers how airplanes could help them fight better on the ground. To further build up his credibility, Arnold entered aviation competitions, winning the coveted McKay flying trophy. He also earned the first aviator's badge awarded to anyone in the United States military, set a world altitude record of 6,540 feet, and earned an expert aviator's certificate. After World War I, and during which he had served in the infantry, Arnold was sent to the army's prestigious Command and General Staff School, where he performed well and was promoted to the rank of lieutenant colonel. His air-oriented accomplishments, coupled with his strong background in traditional army strategies and tactics, added to Arnold's credibility.

Arnold continued to work against tradition, organizational barriers, and skeptical colleagues to make the Army Air Corps a full-fledged, independent arm of the United States military. Little by little he continued to win small victories. The army's key decision makers resisted the establishment of an Army Air Corps, but the man pushing the idea most persistently—Hap Arnold—was one of them, so they had to listen. He was a West Point graduate, had served effectively in the infantry during World War I, and had graduated from the elite Command and General Staff School. Arnold was not an outsider trying to sell insiders on a new and innovative concept; he was an insider—one of *them*. He thus had sufficient credibility to gain access to the decision makers and convince them to listen to his theories.

As World War II approached, Arnold finally began to make real progress in establishing an independent air arm separate from the army. Then the Japanese attacked Pearl Harbor, and years of resistance were suddenly washed away in a tidal wave of fury against the Japanese. Almost overnight, skeptics became supporters. Clearly, air power would play a key role in World War II—if not the key role. Shortly after the devastating surprise attack at Pearl Harbor, Arnold—who had envisioned an air force almost 40 years earlier—was named commanding general of the Army Air Corps.

Arnold's next move to establish a stand-alone, independent air arm of the U.S. military came during World War II, when he organized the 20th Air Force to conduct a comprehensive and focused bombing campaign against Japan. He was a five-star general when he retired in March 1946. Then, finally—thanks to the perseverance and credibility of Hap Arnold—on September 18, 1947, what had been the air arm of the U.S. Army became a stand-alone branch of the military—the U.S. Air Force.

Chief Sitting Bull[13]

Known to Americans as Chief Sitting Bull, Tatanka Yotanka became head chief of the entire Sioux nation. "The dignity, courage, spiritual presence, and moral force of this Sioux chief and medicine man made him the most famous Indian warrior–leader in American history. A member of the Hunkpapa tribe, a branch of the Teton Sioux, Sitting Bull was born on the Grand River in South Dakota, the son of a renowned chief. He showed leadership qualities from an early age. He was famed as a hunter by the time he was 10 years old, and he distinguished himself on the field of battle by 14."[14] The career of Sitting Bull teaches the following lesson:

It takes credibility to hold an organization together during tough times.

Sitting Bull quickly established himself as not just a brave warrior but also an outstanding leader. He led both major engagements and minor skirmishes from 1862 through 1876 with excellent results. The culmination of his fighting years occurred at the Battle of the Little Big Horn, remembered in American folklore as Custer's Last Stand. Sitting Bull did not lead this engagement on the battlefield. Rather, he was the chief of the war council that brought together the Sioux, Cheyenne, and Arapaho and made the engagement possible. He was also the medicine man who "blessed" the Indians going into battle to make their victory possible.

As great a military leader as he was, Sitting Bull's accomplishments off the battlefield were even greater. The Battle of the Little Big Horn was a major victory for the western Indian tribes, but it would be their last. From that day forward, the tribes were pursued relentlessly by the United States Army. In addition, gold discovered in the lands they claimed as home produced an onslaught of settlement that could not be resisted. This is when Sitting Bull's most important leadership qualities became apparent. Almost single-handedly, the old chief held his people together, leading them through terrible times of hunger, depredation, and disease.

Later, when his tribes had been placed on reservations and were expected to learn the ways of their overseers, Sitting Bull worked hard to help his people maintain their traditional Sioux culture. "Sitting Bull commanded the respect and even awe not only of the Sioux he led, but of other Indians and many whites. He acquired his reputation through courageous and wise performance, then maintained his leadership position by offering a continual example of steadfast courage and a refusal to yield."[15]

To keep them from starving to death, Sitting Bull had to convince his proud warriors and their families to do the unthinkable—surrender to the hated and feared white man. Yet, while on the reservation, he had to resist the incursion of the white man's culture into that of his Indian people. He was able to do both for as long as he did because he had credibility in both camps. He could tell his starving warriors to surrender because none among them was his equal as a warrior. He could tell his captors to treat his people with dignity and respect because they admired him as a brave and resourceful adversary. Sitting Bull was able to lead his people through the most difficult period in their lives because he had credibility.

Francis Turner[16]

Few people who travel by car today can remember when there were no interstate highways in the United States. We use interstate highways so frequently that it seems as if they were always there, but they weren't. And without the leadership of a little-known engineer and bureaucrat named Francis Turner, they might never have been. The career of Francis Turner teaches the following lesson:

> To maintain their credibility, leaders must resist the temptation to play favorites.

Turner was not the visionary leader who decided the United States needed a system of interstate highways; that person was President Dwight D. Eisenhower. While serving as Supreme Allied Commander in Europe during World War II, Eisenhower was able to see firsthand how quickly and efficiently Hitler's military commanders could move tanks, trucks, and mobile artillery on its nationwide system of autobahns. The German autobahns made such an impression on Eisenhower that once he was elected president of the United States, he made creating a national system of wide, high-speed highways a top priority. The engineer chosen to create this system was Francis Turner.

Turner helped draft the legislation Eisenhower pushed through Congress to create the largest public works project in the history of the country. It was also one of the most controversial projects. In fact, it is doubtful that a president of lesser stature than Eisenhower could have convinced the American public to support the project. It is also doubtful that any engineer of less credibility than Turner could have completed the project.

The project was highly controversial for three reasons principally. First, although the federal government would underwrite most of the land acquisition and construction costs, the taxpayers in each state would have to provide a percentage of the funding for the interstate highways that ran through their state. Even just a fraction of the cost of constructing an interstate highway was huge. Many of the states were concerned they would not have sufficient funds to pay their percentage. Second, an interstate highway system would require a lot of land—land owned by private citizens who did not necessarily want to give it up. Third, although every effort would be made to avoid populated areas, there would still be American citizens whose homes were located right in the path of an interstate highway. These citizens would have to be uprooted— some of them from homes that had been in their families for generations.

The credibility of Francis Turner was built on his technical and bureaucratic skills coupled with his integrity. One story about Turner's integrity shows why he had the credibility to lead a project that generated so much controversy and disruption in the lives of so many people. Turner's parents lived in Fort Worth, Texas. One morning they awoke to find a survey flag right in the middle of their front yard. It marked the route of the new interstate highway through Fort Worth. Turner's parents pleaded with him to use his authority to save their home.

As the man in charge, Turner could have rerouted the Fort Worth highway and saved his parent's home. But knowing that thousands of other American families across the country were waking up to find survey markers in their yards, too, Turner told his parents that although he could change the route, he wouldn't. Knowledge that Turner would not afford even his own parents special treatment in the process gave him the credibility he needed to see the controversial project through to completion.

Margaret "Meg" Whitman[17]

During a time in which once promising dot.com companies were falling out of favor on Wall Street faster than cottage cheese goes bad, computer auction company eBay not only held its own but prospered. Much of the credit goes to CEO Margaret "Meg" Whitman. She leads a company with an economy larger than that of Iceland and is considered by business insiders to be the most successful of all Internet executives. The career of Meg Whitman teaches this lesson:

Substance is more important than image in maintaining credibility.

While other Internet companies were either struggling to survive or actually going out of business, Meg Whitman led eBay to unprecedented profits by keeping her people focused on the company's core competencies. Whitman knew intuitively what made her company work—not the products sold, but providing a convenient and efficient marketplace for the selling and buying of those products.

Meg Whitman's credibility with stakeholders—employees, shareholders, sellers, and buyers—comes from her solid credentials (competence) and her low-key style (substance over image). Whitman holds an MBA from Harvard and had already established a solid and successful record in business before joining eBay. She worked in brand management at Procter & Gamble; was an executive at Disney, for whom she opened the first Disney stores in Japan; helped revive the Keds brand at Stride Rite shoe company, and rejuvenated Florist's Transworld Delivery (FTD). Her success in these challenging positions gave her the credibility to land the job as CEO at eBay.

In addition to her demonstrated competence, Whitman has a low-key, down-to-earth approach to leadership that employees find nonthreatening.

In spite of being widely recognized as one of the most powerful women in America, in spite of owning more than $600 million in eBay stock, and in spite of being one of the most successful CEOs—male or female—in the world of business, Whitman still works out of a large cubicle, dresses casually, and likes to answer her own email. Her down-to-earth leadership style adds to her credibility by showing eBay's employees that, even with all of her success, Meg Whitman is still someone they can relate to.

Summary

1. Credibility is the ability to inspire belief. Leadership characteristics that inspire and maintain credibility are honesty, ability to be forward-looking, competence, and ability to inspire.

2. Credibility is important to leaders because people are reluctant to follow those they do not believe in. People who believe in their leader take pride in their organization, exhibit team spirit, believe their values match those of the organization, feel they are a part of the organization, and make a commitment to the organization.

3. To establish credibility, leaders must have strong moral character; positive, trusting, mutually supportive relationships; industry knowledge; intuition; solid experience; past success; and field-specific ability. To maintain credibility, leaders must set a positive example, support their people, admit mistakes, and follow through on promises.

4. Stewardship means leading in such a way as to ensure that the organization will be in better condition when you leave than when you started. Leaders who are good stewards take care of the resources entrusted to them and use them wisely, effectively, and efficiently.

5. The career of Henry Harley "Hap" Arnold teaches that with credibility and perseverance a leader can achieve a lofty vision.

6. The life of Chief Sitting Bull teaches that it takes credibility to hold an organization together during tough times.

7. The career of Francis Turner teaches that in order to maintain their credibility, leaders must resist the temptation to play favorites.

8. The career of Meg Whitman teaches that substance is more important than image to a leader in maintaining credibility.

Key Terms and Concepts

Ability	Admit mistakes
Ability to be forward-looking	Character

Competence	Knowledge
Credibility	Past success
Experience	Relationships
Follow through	Set the example
Honesty	Stewardship
Inspiration	Support your people
Intuition	

Review Questions

1. Discuss credibility as it relates to leadership.
2. Explain why credibility is so important to those who would lead.
3. How can a leader establish credibility with those he wants to lead?
4. How can a leader maintain credibility with his followers?
5. Discuss stewardship as it relates to credibility.
6. Explain the credibility lesson taught by the career of Hap Arnold.
7. Explain the credibility lesson taught by the life of Chief Sitting Bull.
8. Explain the credibility lesson taught by the career of Francis Turner.
9. Explain the credibility lesson taught by the career of Meg Whitman.

LEADERSHIP SIMULATION CASES

The following simulations are provided to generate additional thought and discussion about the principles of leadership explained in this chapter. Readers are encouraged to consider how the situations presented in these cases might apply to them and to discuss the cases with other leaders and leadership candidates.

CASE 3.1 How Can I Establish Credibility with My Team?

Mona Goodsen had more than 15 years of solid experience in the field of chemical engineering. Unfortunately, none of it was with the company whose chemical engineering department she had just been hired to lead. As she assessed her situation, Goodsen knew instinctively that credibility would be an issue. First, the director of chemical engineering she was replacing had been with the company for 20 years and was hugely popular with employees. Second, most of her employees were men who had never reported to a woman. Third, her previous employer had been a competitor of her new employer. This final point was the one that worried Goodsen the most.

Wanting to get off to a positive start, Goodsen decided to ask the advice of a friend who had been through a similar experience. They met for lunch, and Goodsen got right to the point: "What can I do to establish credibility with my new direct reports?"

Discussion Questions

1. Assess Mona Goodsen's situation. What problems do you think she might encounter trying to establish credibility in her new position?
2. How would you answer the question Goodsen posed to her friend?

CASE 3.2 How Can I Maintain My Credibility?

John Parker came to his new position as CEO of Innovative Technologies, Inc. (ITI), with a high level of credibility. He was well known in the engineering and manufacturing world as the entrepreneur who had invented several innovative products that make life more comfortable for business-people whose jobs involve frequent flying. The customer niche ITI wants to reach uses the types of products marketed through the magazines and catalogs found in the seat backs of major airlines. Clearly, John Parker was just the man to help ITI expand its product line for this market.

But John Parker knew that credibility could be a fleeting phenomenon. If he was going to achieve the vision he had established for ITI, he would have to maintain his credibility with employees, shareholders, customers, catalog partners, and Wall Street brokers. He faced a daunting task. As he sat alone in his new office, Parker took out a pen and a legal pad. At the top of it he wrote the following question: "How can I maintain my credibility with all stakeholders?"

Discussion Questions

1. John Parker brought established credibility to his new job. What problems do you think he might encounter maintaining his credibility?
2. How would you answer the question Parker wrote on his legal pad?

CASE 3.3 A Case of Poor Stewardship

Mike Adams was incensed over an article that appeared on the business page of the prestigious *New England Gazette*. The headline of the article read,

"Adams Called Poor Steward." The article went on to claim that in spite of his hard-charging, no-nonsense business deals—deals that had turned his company around—Mike Adams was a poor steward who would eventually "wreck" the company "if he stayed at the helm long enough."

The reporter claimed to have interviewed several key employees off the record who were concerned that Adams "rode employees hard and then tossed them aside like yesterday's newspaper, . . . put off upgrading technologies to make the bottom line look better in the short run but was undercutting the company's manufacturing capability in the long run," and was "making inflated claims to customers that the company's products could not live up to."

Discussion Questions

1. Have you ever worked with or for an individual who was a poor steward of the company's resources? If so, what problems did this individual's poor stewardship cause?

2. If the quotes in the newspaper article are true, what do you think of the long-term prospects for Mike Adams and his company?

Endnotes

[1] Louis E. Boone, *Quotable Business*, 2nd ed. (New York: Random House, 1999), 102.

[2] James M. Kouzes and Barry Z. Posner, *The Leadership Challenge* (San Francisco: Jossey-Bass, 2002), 32–33.

[3] *American Heritage College Dictionary*, 3rd ed. (New York: Houghton Mifflin, 1993), 325.

[4] Kouzes and Posner, 24–25.

[5] Kouzes and Posner, 33.

[6] Kouzes and Posner, 33.

[7] Louis E. Boone, 103.

[8] John C. Maxwell, *The 21 Irrefutable Laws of Leadership* (Nashville, TN: Thomas Nelson, 1998), 50–51.

[9] Thomas J. Neff and James M. Citrin, *Lessons from the Top* (New York: Currency & Doubleday, 2001), 99–103.

[10] Louis E. Boone, 196.

[11] Peter Block, *Stewardship: Choosing Service over Self-Interest* (San Francisco: Berrett–Koehler, 1996), xx.

[12] Alan Axelrod, *Profiles in Leadership* (Upper Saddle River, NJ: Prentice Hall, 2003), 40–42.

[13] Alan Axelrod, 496–498.

[14] Alan Axelrod, 496.

[15] Alan Axelrod, 497.

[16] Lewis Lord, "The Superhighway Builder," *U.S. News & World Report*, Special Collector's Edition, 2002, 74.

[17] Loren Fox, *"Meg Whitman,"* salon.com, http://dir.salon.com/people/bc/2001/11/27/whitman/index.html.

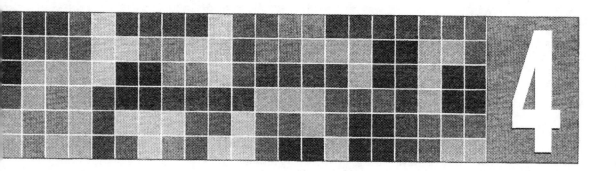

Develop a Can-Do Attitude and Seek Responsibility

"Victorious leaders feel the alternative to winning is totally unacceptable, so they figure out what must be done to achieve victory, and they go after it with everything at their disposal."[1]

—John Maxwell
Leadership Expert

OBJECTIVES

- Discuss the concept of the can-do attitude in the context of leadership.
- Explain why a can-do attitude is important to leaders.
- Describe the leader's responsibilities in maintaining a can-do attitude.
- Explain how the attitudes of leaders affect those of their followers.
- Explain how a leader can develop a can-do attitude.
- Describe the leader's responsibility for providing calm in the storm.
- Explain why it is important for leaders to take charge and seek responsibility.
- Summarize the can-do lessons taught by selected effective leaders.

Winston Churchill was prime minister of Great Britain for the first time during the difficult years of World War II. He is viewed by students of leadership as the man who almost single-handedly held his country together during the difficult early days of World War II when Great Britain stood virtually alone against Adolf Hitler's seemingly unstoppable Nazi juggernaut. More than anything else, it was Churchill's can-do attitude that bolstered the resolve of his beleaguered countrymen.

"The key to Churchill's courage was his unbounded optimism. Only an optimist can be courageous, because courage depends on hopefulness that dangers and hazards can be overcome. . . . He [Churchill] deprecated negative thinking."[2] The positive effect of Churchill's can-do attitude can be seen in the comments of one of his contemporaries: "We owed a good deal in those early days to the courage and inspiration of Winston Churchill who, undaunted by difficulties and losses, set an infectious example to those of his colleagues who had given less thought than he, if indeed any thought at all, to war problems. . . . His stout attitude did something to hearten his colleagues."[3]

At a time when his countrymen needed it most, Churchill adopted a can-do attitude and took responsibility for leading the defense of Great Britain. His is an excellent example for technical professionals who hope to lead organizations in the battle of the marketplace. The can-do attitude Churchill used to help his country eventually win in a global war is the same attitude you need to help your company win in the global marketplace.

WHAT IS A CAN-DO ATTITUDE?

A can-do attitude is an outward manifestation of an inner conviction that asserts, "Whatever the job, I can get it done; whatever the challenge, I can meet it; whatever the obstacle, I can overcome it." Such an optimistic attitude should not be confused with false bravado or unwarranted optimism. Rather, a can-do attitude is founded in the leader's conviction that, within the bounds of legality and ethics, she can and will do what is necessary to succeed. A can-do attitude has the following elements: optimism, initiative, determination, responsibility, and accountability.

The story of Herb Kelleher and Southwest Airlines demonstrates the value of a can-do attitude. Kelleher, an attorney, is one of the three founders of Southwest Airlines. Now one of the most successful airlines in the United States, Southwest almost didn't get off the ground. Before the ink was dry on its incorporation papers, Southwest Airlines was ganged up on by competing airlines that tried to keep it on the ground with a series of expensive lawsuits. "One court battle followed another, and one man, more than any other, made the fight his own: Herb Kelleher. When their startup capital was gone, and they seemed to be defeated, the board wanted to give up. However, Kelleher said, 'Let's go one more round with them. I will continue to represent the company in court, and I'll postpone any legal fees and pay every cent of the court costs out of my own pocket.' Finally when their case made it to the Texas Supreme Court, they won, and they were at last able to put their planes in the air."[4]

The can-do attitude of Herb Kelleher helped him lead Southwest Airlines through many more struggles until it eventually grew from a company with four airplanes and total assets of $22 million to a successful airline with almost 275 airplanes and total assets exceeding $4 billion.[5] Kelleher's example shows the value of a can-do attitude in leading organizations.

WHY IS A CAN-DO ATTITUDE IMPORTANT?

The attitude of a leader is often mirrored by his followers. A leader with a bad attitude is likely to spawn bad attitudes in his followers; if, that is, he can manage to maintain any followers. People do not like to follow someone with a bad attitude. In contrast, as the earlier example of Winston Churchill clearly shows, a leader with an optimistic, can-do attitude can sustain the morale of his followers through even the most difficult times. "Optimism is also the key to the can-do spirit, to the don't-take-no-for-an-answer attitude that is essential to successful executive leadership. Nearly all human organizations are subject to an inertia that results in an it-can't-be-done attitude. This was always unacceptable to Churchill."[6] Such an attitude must also be unacceptable to technical professionals who want to lead.

LEADERS ARE RESPONSIBLE FOR THEIR ATTITUDE

There is a tendency among some to think that an optimistic, can-do attitude is a gift people either have or don't have. Such people see a can-do attitude as something you are born with or without. Nothing could be further from the truth. At the beginning of this chapter, Winston Churchill was used as an example of a leader who exemplified the can-do attitude. Churchill is such a good example precisely because he had to work so hard to maintain the can-do attitude for which he is known. What is less known, except by careful students of history, is that Churchill suffered from frequent and severe bouts of depression. His well-known can-do attitude was a hard-won attribute that he had to work continually to maintain. It is the same with most leaders.

A can-do attitude is as much a tool to a leader as a hammer is to a carpenter, and just as the carpenter is responsible for maintaining his tools, the leader is responsible for maintaining his attitude. The following quote is often used to remind people that a positive, can-do attitude is a choice, not a gift or an accident of birth: "We cannot choose how many years we will live, but we can choose how much life those years will have. We cannot control the beauty of our face, but we can control the expression on it. We cannot control life's difficult moments, but we can choose to make life less difficult. We cannot control the negative atmosphere of the world, but we can control the atmosphere of our minds. Too often, we try to choose to control things we cannot. Too seldom, we choose to control what we can . . . our attitude."[7]

LEADERS AFFECT THE ATTITUDES OF THEIR FOLLOWERS

Consider the following words of John C. Maxwell: "Leadership is influence. People catch our attitudes just like they catch our colds—by getting close to us. One of the most gripping thoughts to ever enter my mind centers on my influence as a leader. It is important that I possess a great attitude, not only for my own success, but also for the benefit of others. My responsibilities as a leader must always be viewed in light of the many, not just myself."[8]

The fact that followers will look to their leaders for the type of attitude they should adopt is precisely why it is so important for leaders to maintain a positive attitude. In the foregoing quote, Maxwell makes the point that an attitude spreads in the same way as a cold. This is an apt description of what actually occurs. When one person in an office gets a cold, it seems that before long everybody has one. The same can be said of a bad attitude. Furthermore, just as it is easier to catch a cold than to prevent one, it is easier to spread a bad attitude than prevent one. This is why it is so important for leaders to be both persistent and consistent in displaying a can-do attitude.

HOW TO DEVELOP A CAN-DO ATTITUDE

Some people might be born with a can-do attitude, but most are not. Developing an optimistic attitude is like developing a muscle: (1) it takes hard work, determination, and constant effort; and (2) if you stop working at it, you can quickly lose it. A process for developing an optimistic, can-do attitude is shown in Figure 4.1. Each step is explained in the following discussion.

Leadership Tip

"Whenever you are asked if you can do a job, tell 'em, Certainly, I can—then get busy and find out how to do it."[9]

—Theodore Roosevelt
American President

> **Four-Step Method for
> Developing a Can-Do Attitude**
>
> 1. *Assess:* Identify your weaknesses (e.g., negative thoughts, behaviors, or feelings).
> 2. *Plan:* Develop a plan for overcoming your negative attitudes and replace them with positive ones.
> 3. *Implement:* Put your plan into action.
> 4. *Monitor and Adjust:* Determine whether the plan is working. Make adjustments as necessary.

FIGURE 4.1   Process for developing an optimistic, can-do attitude.

Assess

Self-assessment is always difficult and sometimes painful. People do not like to focus too much attention on their personal weaknesses. However, the ability to look at yourself objectively, identify problems, and do what is necessary to correct them is one essential mark of a mature professional who has the potential to be an effective leader. When assessing your attitude, identify any major weaknesses in your thoughts, feelings, and behavior.

The results of the self-assessment conducted in this step are the beginning point for the plan you will develop in the next step. Figure 4.2 is a self-assessment checklist technical professionals can use to begin the process. Use this instrument to get started, but do not be limited by it. If you think you have attitudinal weaknesses that do not appear in the instrument, write them down. Figure 4.3 shows the results of a hypothetical self-analysis performed using the checklist shown in Figure 4.2.

Plan

Your plan should have the following elements: a vision, broad goals based on the self-assessment, specific actions to be taken to achieve the goals, and metrics for measuring progress. Figure 4.4 is a portion of an improvement plan that was developed based on the results of the self-assessment shown in Figure 4.3. Note that the plan, although just partial, contains all of the necessary elements—a vision, goals, action steps, and metrics for measuring progress.

■ *Vision.*   Imagine yourself as you would like to be. Create a mental image of yourself as a leader with a can-do attitude, and paint a "word picture" of that image. This word picture is your vision as it relates to a can-do attitude. Refer to the vision statement in Figure 4.4 for an example.

Checklist for an Attitudinal Self-Assessment		
Yes	**No**	
❏	❏	1. Are your thoughts about people generally positive?
❏	❏	2. Are your thoughts about your work generally positive?
❏	❏	3. Can you typically disagree with people without being disagreeable?
❏	❏	4. Do you typically maintain your composure when under pressure?
❏	❏	5. Do you typically perform well when under stress?
❏	❏	6. In most situations do you feel like you can get the job done?
❏	❏	7. Are you able to delegate work to people who do things differently from you?
❏	❏	8. Do you typically seek responsibility instead of waiting for it to be assigned?
❏	❏	9. Are you typically positive about finding solutions when unexpected problems arise?
❏	❏	10. Does your behavior encourage perseverance when obstacles stand in the way of success?

FIGURE 4.2   Self-assessment can be painful but is a necessary part of self-improvement.

■ *Goals.* Convert the weaknesses identified during the self-assessment into goals. State the goals in behavioral (doing or action) terms as an improvement you would like to make. For example, in Figure 4.3 one of the hypothetical weaknesses is the following:

> I don't like it when people disagree with me. Consequently, I often respond by being disagreeable.

In Figure 4.4 this problem area was converted into a behaviorally stated goal:

> Learn to disagree with people without being disagreeable.

Sample Results of Attitudinal Self-Assessment

- I don't like it when people disagree with me. Consequently, I often respond by being disagreeable.
- I sometimes lose my composure when under pressure.
- I try to do too much myself because I don't like the way others do the work.
- I become discouraged too quickly when unexpected problems arise.

FIGURE 4.3 Identifying attitudinal problems is the first step toward correcting them.

By converting the attitudinal problem into a behaviorally stated goal, you encourage improvement in two ways. First, you eliminate all ambiguity. Having written down the goal, there is now no question of what you need to do in order to improve. Second, you build in personal accountability by creating an expectation—in writing—that can be measured to determine whether progress is being made.

■ *Action steps.* Identify specific action steps you can take to accomplish your goal. Action steps identify more clearly than the goal exactly what must be done. As shown in Figure 4.4, each goal has its own unique set of action steps. Each action step, if properly taken, should move you closer to achievement of the goal in question.

■ *Metrics for measuring progress.* There is a management principle that says, "if you want to make progress, measure it." Measurement is a necessary ingredient in the recipe for accountability. This is why people who are trying to lose weight are required to weigh themselves daily. If your action steps are stated in behavioral terms, they can be measured. This is important because you will make better progress if you invest the time and effort to measure results.

You might question the need to develop a written plan for improving your can-do attitude. You might tell yourself, "Now that I've completed the self-assessment, I know what needs to be done. I don't need to write it down." Unfortunately, experience shows that those who apply this rationale rarely succeed. Experience shows that those who are willing to invest the time and effort to develop a written plan are more likely to also invest the time and effort necessary to make improvements.

Partial Plan for Improvement to Create a Can-Do Attitude

Vision

To have a positive, can-do attitude that inspires confidence in my followers.

Goals with Action Steps

1. Learn to disagree with people without being disagreeable.
   a. When I disagree with someone, I will silently count to myself until the anger impulse passes. If it does not pass, I will not speak.
   b. Make a list of tactful ways to indicate disagreement with people (e.g., "I see your point. Would you be open to hearing a different opinion?" or "Interesting idea, John, but may I suggest another strategy we might consider?").
   c. Before meeting with people, practice the responses I wrote down in the previous action steps.
   d. When I feel my temper flaring during a discussion, I will repeat the following statement to myself until the impulse subsides: "In a discussion, if I lose my temper, I lose *period.*"

2. Learn to maintain my composure when under pressure.
   a. In high-pressure situations, I will practice concentrating on the goal instead of the clock.
   b. When I feel the pressure building, I will take three deep breaths and repeat the following statement to myself until I calm down: "Calm, steady, and deliberate gets the job done."

Metrics for Measuring Progress

1. Keep a daily log for a month, recording the number of instances of disagreements I encounter each day, followed by the number of times I successfully remained silent until the anger impulse passed.

2. Keep a record in my daily log of which tactful statements I used to indicate my disagreement with someone. Also indicate how often I used each statement.

3. Keep a tally of the number of times I successfully calmed myself in high-pressure situations.

FIGURE 4.4   Developing a can-do attitude is essential for self-improvement.

Implement, Monitor, and Adjust

Once your plan is complete, implement the plan and use the metrics included in it to measure progress. If you are making acceptable progress, continue on course. If not, make adjustments. A plan is just that—a plan. If it is working, stay with it. If not, drop the ineffective strategies and try new ones. An old saying credited to the Apache Indians applies here: If the horse you are riding dies, climb off and get on another. In other words, do not become wedded to your plan. If the plan isn't working, revise it. It is the progress toward improvement that matters, not the plan.

LEADERS PROVIDE A SENSE OF CALM IN THE MIDDLE OF THE STORM

The modern marketplace is intensely fast-paced and competitive. It has a way of imposing deadlines without asking (or caring) if they are realistic. In today's helter-skelter workplace, when a customer asks for something on one day, the customer often wants it delivered *yesterday*. Consequently, technical professionals often find themselves with too much work and too little time. Those who hope to be leaders must develop the ability to stay calm in even the most hectic circumstances. This is important because followers who see their leader harried, addled, and frazzled will soon become that way themselves. Agitation can quickly spread, and its effect on work can be decidedly negative.

Winston Churchill was a leader who made a point of providing his followers with a calm example during hectic times. "Churchill was no stranger to the number one problem faced by executives—stress. Churchill's colleagues and friends marveled at how calm he was amid the most trying circumstances. He would, for example, set up his painting easel near the front line trenches in World War I and paint as shells were exploding nearby. The lesson of Churchill's extraordinary calm and aversion to haste is that hastiness dilutes your concentration, disrupts your priorities, and makes it impossible to follow a consistent method of work."[10]

People who become agitated when the pressure is on do so because they focus on the problem, not the goal. They begin to concern themselves more with the potentially negative consequences of failure than with what must be done to succeed. This is like the tennis player who is so worried about losing the match that he focuses on the scoreboard rather than the ball. The key to staying calm during turbulent times is to focus on the job to be done, not the deadline and not the results of missing the deadline. In hectic situations, leaders must be especially attentive to staying focused, staying calm, and helping their followers stay calm.

Leadership Profile   David Johnson's Can-Do Attitude at Campbell Soup[11]

David Johnson's plan when he took over as CEO of Campbell Soup Company was simple. He planned to overtake every competitor in the soup-making business, and with his positive, optimistic, can-do leadership, that is precisely what Campbell Soup Company did. According to Johnson, "If you want to be a world-class performer, and I don't want to have anybody with me who doesn't dream of that, then you're not going to set goals that are easy to achieve. You won't do that, because you wouldn't be able to live with yourself. You would not be worth knowing."[12]

Johnson's success at Campbell Soup was astounding. Using his formula for success, 20-20-20 (20 percent annual earnings growth, 20 percent return on equity, and 20 percent return on invested cash), Johnson turned the soup maker into a profit-making powerhouse. Under his can-do leadership, Campbell Soup gained an 85 percent share of the condensed soup market in the United States and began expanding globally. He streamlined the company by eliminating layers of corporate structure, selling off poorly performing businesses, and canceling nonperforming product lines. Using the money saved by these cost-cutting strategies, Johnson began buying companies he thought had good potential. As a result of wise acquisitions, Campbell Soup Company is now the parent of Pace Mexican (salsa and sauces), Prego (spaghetti sauce), Pepperidge Farm and Arnotts (baked goods), and Godiva chocolates.

Numerous executives at Campbell's became wealthy during Johnson's tenure as CEO. He set high goals and had high expectations, but for those who met their goals and his expectations, the financial rewards could be immense. Corporate bonuses and stock options were liberally awarded, but only as a result of excellent performance and in amounts commensurate with performance. In fact, during Johnson's tenure as CEO, only about 20 percent of an executive's total compensation came in the form of salary. The largest percentage of the compensation package came in the form of bonuses and stock options. People in leadership positions at Campbell Soup Company who performed well profited in direct proportion to the company's profits and their individual performance in generating those profits. Under Johnson, leaders who failed to perform were given a second chance. Those who achieved their goals the second time around were rewarded accordingly. Those who did not were given the opportunity to pursue another line of work.

The one constant that more than anything else drove the performance of Campbell Soup Company for a decade was the can-do attitude of David Johnson. He encouraged people in leadership positions to set high goals, but he gave them the support they needed to achieve the goals. He was always there pushing, pulling, and telling people "you can do it." Because his can-do attitude was contagious, most did. As a result, their performance and that of the company just got better and better every year.

CAN-DO LEADERS TAKE CHARGE AND SEEK RESPONSIBILITY

Can-do leadership involves more than just inspiring optimism in followers. Can-do leaders are people who, when they see something that needs doing, take charge of the situation and get it done. In addition, they take responsibility for their actions and expect to be held accountable. By this you can see that the element of risk is always present in leadership. This risk factor is just one more reason why so few people ever become effective leaders. After all, in any situation, the surest way to make no mistakes is to do nothing at all. Can-do leaders are unwilling to do nothing when it is clear that something needs to be done. Instead, they take the initiative, take charge, and seek responsibility. Can-do leaders never sit back and wait to be told what to do, nor do they shrink from being held accountable for their actions.

According to Kouzes and Posner, "Leaders seize the initiative with enthusiasm, determination, and a desire to make something happen. They embrace the challenges presented by the shifts in their industries or the new demands of the marketplace. They commit themselves to creating new possibilities that make a meaningful difference. . . . Leadership bests are filled with stress. . . . But instead of being debilitated by the stress of a difficult experience they [can-do leaders] are challenged and energized by it. Stress always accompanies the pursuit of excellence.[13]

Can-do leaders are people who, when asked to move a mountain, will find a way to accomplish the task. Winston Churchill was just such a leader. "Though Churchill never moved a mountain, he did once move a small river, during his schooldays as a young cadet at Sandhurst. He dropped a gold watch his father had given him into a deep pool in a stream. Despite attempts to dive to the bottom of the pool, he was unable to retrieve the watch. Dredging was equally unsuccessful. So Churchill hired 23 soldiers from a nearby infantry attachment to dig a new channel for the river, after which Churchill procured a fire engine to pump out the pool. The watch was recovered."[14]

Leadership Tip

"Leaders make something happen. . . . they are proactive and able to make something happen under conditions of extreme uncertainty and urgency. In fact, leadership is needed more during times of uncertainty than in times of stability."[15]

—James M. Kouzes and Barry Z. Posner
Leadership Researchers and Authors

Winston Churchill exemplified all of the concepts that together make up a can-do attitude: optimism, initiative, determination, responsibility, and accountability. Churchill loved his father above all else, and he knew his father valued the watch he had given his son. When the watch appeared to be irretrievably lost, many young men would have given it up as a lost cause, but Churchill was cut from cloth different from that of most young men.

Rather than despair and give up, young Churchill maintained an *optimistic outlook* that said, "I can solve this problem." After several attempts to retrieve the watch by diving into the pool, Churchill saw that he would need a new and different strategy. Rather than give up, he took the *initiative* and hired a team of soldiers to help him divert the stream. With that done, there was still the problem of the pool—a pool so deep that diving still had to be ruled out as an option.

Here Churchill showed the *determination* of a can-do leader. He procured a fire truck and pumped the pool dry. Finally, throughout the entire project, Churchill exemplified the traits of *responsibility* and *accountability*. He not only took responsibility for getting the watch out of the pool but also expected to be held accountable for his actions in doing so. Churchill was accountable to his father for retrieving the watch. He was accountable to the soldiers who helped him divert the river for a certain amount of financial remuneration. They were well paid. He was accountable to the fire department for properly using and returning its pumping truck. He returned the truck in good working order. Finally, he was accountable to the community for returning the diverted river to its normal path, which he did. The point here, and it is a critical point, is that the leader is responsible for both getting the job done and how he gets it done.

LESSONS ON CAN-DO ATTITUDE FROM SELECTED LEADERS

Following are excerpts from the lives of several leaders that exemplify some of the can-do attitude principles set forth in this chapter. The leaders selected for inclusion are Jean Campbell, Harry Truman, John F. Kennedy, Abraham Lincoln, and Booker T. Washington.

Jean Campbell[16]

One of the hallmark traits of a leader with a can-do attitude is initiative. When others let adversity befuddle them into inactivity or frighten them into surrender, the can-do leader takes the initiative and persists in finding a way to get the job done. Jean Campbell, founder of Synergistic Systems, Inc. (SSI)—a computer-based medical billing company—proved to be such a leader during the most extreme crisis she or her company had ever faced. The can-do attitude lesson of Jean Campbell is as follows:

A leader who is willing to take the initiative can overcome enormous obstacles and turn even a disaster into a success.

Early one morning while many residents still slept, the peaceful tranquility of California's San Fernando Valley was violently rocked by a devastating earthquake. SSI's corporate facilities were located close to the hardest-hit area in the valley. Rushing through the rubble, not yet knowing if her facility had even survived, Campbell braced herself. Even so, she was shocked to find a building that had simply folded in on itself and collapsed. The situation was even worse than she had feared. SSI had no facility from which to operate.

Having found what Jean Campbell found, many people would have given up. But Jean Campbell is not like most people. She did not even hesitate. Rather, Campbell made up her mind that she would (1) keep revenue flowing to her clients, (2) keep paychecks flowing to her employees, and (3) be fully operational within two weeks. These were challenging—some would say unrealistic—goals for a CEO whose corporate headquarters was little more than a shattered pile of rubble.

"Working closely with IBM Business Recovery Services, Jean organized, planned, listened, reassured, and motivated SSI employees and contractors to restore essential services within forty-eight hours and full service in less than ten business days. Jean and her team seized the initiative and energized a partnership of employees, suppliers, and customers so powerful that it overcame the forces of devastation unleashed by nature. She used her initiative and encouraged others to do the same; in so doing, she accomplished the extraordinary amid incredible chaos and change."[17]

In keeping her company operating in the midst of chaos and destruction, Jean Campbell displayed all of the characteristic elements of a can-do attitude. It was her *optimism* that convinced her employees, IBM officials, and customers that by working together they could struggle through. It was her *initiative* in finding ways to get computers back on line and employees operating them that allowed the work of her company to go on in spite of the devastation that was everywhere. It was her *determination* to take care of her customers and employees that gave Campbell the credibility she needed to win the cooperation of others critical to the success of her restoration efforts. Throughout the crisis, everyone involved knew that Jean Campbell took full *responsibility* for getting her company back online and expected to be held *accountable* for producing results, even under circumstances others deemed impossible.

Harry S. Truman[18]

People had underestimated Harry S. Truman all of his life. After all, he didn't look much like a leader. A nondescript, plainspoken man, Truman wore thick spectacles that magnified his eyes and, to most people, looked more like a rural county judge—which he had been—than vice president of the United States. To make matters even worse, plain-looking, plainspoken Truman served as vice president to an enormously popular and charismatic president, Franklin D. Roosevelt. But the men of Battery D of the 129th Field Artillery who served under Truman during World War I could have given the American public a much different and more accurate picture of the vice president. They knew that inside the little man with the thick glasses beat the heart of a capable, courageous, can-do leader. The can-do attitude lesson taught by the career of Harry Truman is:

> A person who suddenly and unexpectedly has leadership thrust upon him can achieve world-changing success if he has determination and the right attitude.

Harry Truman achieved much during his terms as president of the United States. Just one of his achievements—the Berlin Airlift—demonstrates the importance of a can-do attitude. Only a leader with a can-do attitude who simply refused to lose could have accomplished what turned out to be the most challenging but most successful large-scale humanitarian effort in the history of the world.

"Another spectacular Cold War success was the Berlin Airlift of 1948–49, which kept West Berlin supplied with food and fuel after the Soviets blockaded the city in an effort to prevent the creation of an independent, democratic West Germany. The success of the airlift brought about the end of the Berlin blockade and was a major Cold War triumph."[19]

> "The Berlin Airlift was preceded by nearly three years of constant tension and sniping by the Russians against the Western Allies in an attempt to drive them out of West Berlin and West Germany by a variety of tactics, including searches, blocking transportation of even basic supplies to West Berlin and intimidating West German civilians. At midnight on 23rd June, the Russians cut the electrical power to the western sectors of Berlin, and at six o'clock on the 24th of June, they severed all road and barge traffic to and from the city, and at the same time they stopped the transfer of all supplies from the Soviet sector. On 24th June, the western sectors of Berlin were under siege."[20]

To grasp the enormity of what Truman accomplished with the Berlin Airlift, imagine Washington, D.C., having to be supplied with all of its needed food and supplies for just one day. Now multiply that image times three months. The sheer magnitude of the effort boggles the mind. Hundreds of thousands of tons of supplies were needed, and the only way to get them into the city was by cargo planes. During the period of the Berlin Airlift, a cargo plane landed or took off every 3 minutes, 24 hours a day, 7 days a week.

In providing the leadership for the Berlin Airlift, Harry Truman displayed all of the characteristic elements of a can-do attitude. He maintained an air of *optimism* that said, "We can do this" when others were wringing their hands or were on the verge of simply giving in and letting West Berlin fall into the hands of the Soviets. He took the *initiative* and made decisions when others were either afraid to act or were stunned into paralysis by the Soviets' belligerence. He showed *determination* by sticking to his plan and persisting until, months later, the Soviets finally saw the futility of their efforts and gave up. Many of those opposed to the Berlin Airlift were afraid it would lead to

World War III, a nuclear confrontation with the Soviet Union. In pressing ahead with the airlift, Truman took full *responsibility* for the outcome and expected to be held *accountable* for the consequences of his actions.

John F. Kennedy[21]

John F. Kennedy is typically remembered as the youthful president struck down in his prime by an assassin's bullet in Dallas, Texas. He is also remembered for his courageous stand against the Soviet Union during the Cuban missile crisis and for setting America on a course to pursue a bold new social agenda. But one of his brightest moments as a leader came well before he was elected president of the United States. It was as a PT boat skipper during World War II that John F. Kennedy first showed the can-do attitude he later made famous as president. The can-do attitude lesson of John F. Kennedy, PT boat skipper, is as follows:

> With the right attitude, a leader can hold his team together, even in life-and-death situations.

John Kennedy grew up sailing the waters off the coast of Hyannis Port, Massachusetts. Consequently, volunteering for duty on PT boats—although this represented extremely hazardous duty—was a decision that came naturally to the future president. While commanding PT 109 on a war patrol in the south Pacific the night of August 2, 1943, Kennedy and his crew were suddenly thrown violently to the deck as a Japanese destroyer rammed the much smaller American vessel, catching it on fire and splitting the craft in two. Two men were killed immediately. The remaining members of the crew, including Kennedy, leapt into the water to avoid the flames. Kennedy, whose back had been injured during a football game in his college days at Harvard, had to endure intense pain when the collision re-injured his back. In spite of the pain, Kennedy held his crew together, got them to the temporary safety of a nearby island, and eventually—six days later—facilitated their rescue. Kennedy had scratched out a message on a coconut and asked natives to take it to a friendly coast watcher, who in turn notified the American Navy of the location of Kennedy's crew.[22]

In John F. Kennedy's actions following the destruction of his PT boat, one can find all of the characteristic elements of a can-do attitude. Their boat, their only transportation back to safety, was completely destroyed, two crewmen were killed, and others—including Kennedy—were injured. The only land within swimming distance was an island devoid of food and surrounded by Japanese who regularly patrolled the waters off its shores. Kennedy's crew had every reason to despair, but the skipper would have none of it. Throughout the ordeal he remained calm and displayed a positive attitude that said, "We'll get through this."

Once his men were safely on shore, Kennedy did not even wait to rest before he took the *initiative* and began his efforts to have them rescued. The first night of the ordeal, in spite of his injured back, he swam out into the ocean in an attempt to flag down any American vessel that might happen by. He organized his men on the island into shifts so that while some slept, others were awake to be alert to discovery by Japanese patrols. When he and his crew were discovered by natives, he scratched out a message on a coconut, an act that eventually led to their rescue.

Kennedy was *determined* to get his crew safely back into American hands. His determination helped the crew maintain their will to go on in spite of exhaustion, dehydration, and hunger. When Kennedy scratched the rescue note on a coconut, he ran the risk that it might fall into Japanese hands, thereby notifying them of his presence in their territory. However, he took full *responsibility* and expected to be held *accountable* for his actions before, during, and after the destruction of his boat. He was. Following the ordeal, Kennedy was cited for leadership and courage by both the U.S. Navy and the Marine Corps.

Abraham Lincoln[23]

A practice of many can-do leaders is to accept responsibility when things go wrong and give praise where praise is due. This practice reinforces the can-do leader's credibility with followers because it exemplifies two of the characteristic elements of the can-do attitude—*responsibility* and *accountability*. Abraham Lincoln was a master at applying this practice. One of the can-do attitude lessons of Abraham Lincoln's career is as follows:

> There is always an element of risk in being a can-do leader because accepting responsibility and being accountable are risky undertakings. However, the leader who is willing to risk responsibility and accountability can accomplish great things.

"When a subordinate did a good job, Lincoln praised, complimented, and rewarded the individual. On the other hand, he shouldered responsibility when mistakes were made. The president, for example, readily accepted responsibility for the battles lost during the Civil War. He tried to let his generals know that if they failed, he too failed. The loss of the second battle of Bull Run, for example, created a great deal of anger in Washington, most of it directed at Gen. George McClellan because he failed to provide field commander John Pope with appropriate support. It was generally believed at the time that McClellan wanted Pope to fail. As a result, several angered cabinet officers signed a letter of protest condemning McClellan for his conduct during the battle and demanded his dismissal. Lincoln chose instead to appoint McClellan to command the forces in Washington. The cabinet members first heard of the appointment together in session with the president, and an infuriated Secretary of War Stanton exclaimed that no such order had been issued from the War Department. Lincoln then responded somewhat calmly, 'No, Mr. Secretary, the order was mine; and I will be responsible for it to the country.' Lincoln felt that McClellan should not have to bear the entire burden for the loss. . . . So, after the battle, he appointed McClellan at the risk of having his entire cabinet resign."[24]

Years later when President Harry Truman said, "The buck stops here," he was speaking for all leaders, not just presidents of the United States. The "buck" of which he spoke is responsibility coupled with the corresponding account-ability. Can-do leaders, as Lincoln so often showed, must be willing to accept responsibility for the success or failure of the people and organizations they lead, and they must expect to be held accountable for their actions.

Booker T. Washington[25]

Few leaders have more steadfastly exemplified the power of a can-do attitude than Booker T. Washington, the great educator and teacher of self-reliance.

> Booker T. Washington was born a slave in Franklin County, Virginia, in 1856. Raised by his mother, he spent his first nine years on the farm of his master, James Burroughs. After emancipation he moved with his mother and step father to Malden, West Virginia, where he worked as a salt packer and a coal miner, securing a basic education in his spare time. From the principal, Samuel Chapman Armstrong, Washington absorbed an educa-tional philosophy that emphasized the practical training of African Amer-icans in traditional industrial and agricultural occupations. After graduating in 1875 and serving as a teacher at Hampton, Washington ac-cepted an invitation to move to Tuskegee, Alabama, where he opened the Tuskegee Normal and Industrial Institute on 4 July 1881. Through his skill-ful management, extraordinary fund-raising success, and shrewd diplo-macy . . . Washington developed Tuskegee into the most famous African American institution of higher learning in the United States.[26]

The can-do attitude lesson of Booker T. Washington is as follows:

> A leader can overcome even the most difficult obstacles and go on to achieve great deeds if he has a can-do attitude and a lofty vision.

Washington built Tuskegee Institute literally from the ground up. When he arrived in Tuskegee eager to establish a school, what he found was nothing more than acre upon acre of flat, open farmland. There were no classrooms and no dormitories. In fact, there were no buildings of any kind. Although Washington found no buildings, he did find something even better. "Before go-ing to Tuskegee I had expected to find there a building and all the necessary apparatus ready for me to begin teaching. To my disappointment, I found noth-ing of the kind. I did find, though, that which no costly building and appa-ratus can supply—hundreds of hungry, earnest souls who wanted to secure knowledge. . . . But gradually, by patience and hard work, we brought order out of chaos, just as will be true of any problem if we stick to it with patience and wisdom and earnest effort. . . . It means a great deal, I think, to start off on a foundation which one has made for one's self."[27]

Starting off on a foundation one has made for oneself was the basis of Washington's can-do attitude. He began life as a slave with few rights and poor prospects, but through his own hard work and perseverance he became not just an educated man but a great educator of others. Because of this experience, Washington knew he could achieve any goal, handle any problem, and overcome any difficulty he set his mind to. Consequently, he also knew what was necessary to breed that same can-do attitude in others. This is why the earliest students at Tuskegee not only had to study their lessons but also had to build the very classrooms in which they studied and the dormitories in which they lived. They had to raise the food they ate and prepare it for each other. They had to perform all of the daily maintenance duties necessary for the orderly operation of a college. In short, they had to become as self-reliant and self-sufficient as human beings can possibly be. This self-reliance and self-sufficiency was the foundation on which Washington's students developed their own can-do attitudes.

Washington spoke of his approach to instilling a can-do attitude in others in his autobiography: "Few things help an individual more than to place responsibility on him, and to let him know that you trust him. When I have read of labor troubles between employers and employees, I have often thought that many strikes and similar disturbances might be avoided if the employers would cultivate the habit of getting nearer to their employees, of consulting and advising with them, and letting them feel that the interests of the two are the same."[28]

Summary

1. A can-do attitude is an outward manifestation of an inner conviction that asserts, "Whatever the job, I can get it done; whatever the challenge, I can meet it; whatever the obstacle, I can overcome it." Such an optimistic attitude should not be confused with false bravado or unwarranted optimism. A can-do attitude has the following elements: optimism, initiative, determination, responsibility, and accountability.

2. A can-do attitude is important to those who would lead because followers tend to mirror the attitude of their leader. A leader with a bad attitude, if he is able to maintain any followers, will most likely spawn bad attitudes in them. People do not like to follow a person with a bad attitude.

3. A can-do attitude is neither a gift nor something one is born with. Those who display can-do attitudes must first develop them and then maintain them. Maintaining a can-do attitude is similar to maintaining a skill. You must work at it and practice constantly. Leaders who stop working at maintaining their can-do attitude will soon lose it.

4. The four-step model for developing a can-do attitude is as follows: (1) objectively assess yourself and identify any weaknesses in your thoughts, feelings, and behavior; (2) develop a plan for overcoming all weaknesses identified (the plan should contain a vision, goals, action steps, and metrics for measuring progress); (3) implement the plan; and (4) monitor and adjust.

5. It is important for leaders to provide their followers with a sense of calm during hectic times. Followers who see their leaders harried, addled, and frazzled will soon become that way themselves. Such agitation can quickly spread, and its effect on work can be decidedly negative.

6. Can-do leadership involves more than just instilling optimism in followers. Can-do leaders are people who, when they see something that needs to be done, take charge of the situation and get it done. In addition, they expect to be held accountable for their actions.

7. As founder and CEO of Synergistic Systems, Inc., Jean Campbell showed that a leader who is willing to take the initiative can overcome enormous obstacles and turn even a disaster into a success.

8. When Harry S. Truman, upon the death of Franklin D. Roosevelt, became president of the United States, he showed that a person who suddenly and unexpectedly has leadership thrust upon him can achieve world-changing success if he has determination and the right attitude.

9. As the commander of a PT boat rammed by a Japanese destroyer during World War II, John F. Kennedy showed that with the right attitude a leader can hold his team together, even in life-and-death situations.

10. As president of the United States during the Civil War, Abraham Lincoln showed that although there is an element of risk in accepting responsibility and being held accountable, the leader who is willing to take the risk can accomplish great deeds.

11. As the founder and builder of Tuskegee Institute, Booker T. Washington showed that a leader can overcome even the most difficult obstacles and go on to achieve great deeds if he has a lofty vision and a can-do attitude.

Key Terms and Concepts

Action steps

Assess

Calm in the middle of a storm

Can-do attitude

Goals

Implement

Metrics for measuring progress

Monitor and adjust

Plan

Take charge and seek
    responsibility

Vision

Review Questions

1. Explain the concept of the can-do attitude and its various elements.
2. Why is a can-do attitude so important to a leader?
3. What are the leader's responsibilities in maintaining a can-do attitude?
4. Explain how the attitudes of leaders affect those of their followers.
5. Describe the process for developing a can-do attitude.
6. Why is it so important for a leader to provide a sense of calm during hectic times?
7. Explain why it is important for leaders to take charge and seek responsibility.
8. Summarize the lessons taught by the selected effective leaders presented in this chapter.

LEADERSHIP SIMULATION CASES

The following simulations are provided to generate additional thought and discussion about the principles of leadership explained in this chapter. Readers are encouraged to consider how the situations presented in these cases might apply to them and to discuss the cases with other leaders and leadership candidates.

CASE 4.1   Who Needs a Can-Do Attitude?

"I don't buy all this stuff about having a can-do attitude," said Glenda Grason to her fellow vice president Amanda Brown. The two executives with Technical Accessories, Inc. (TAI), were having lunch and discussing a seminar they had just attended. "Why not?" asked Brown. "It made sense to me." "It sounds like nothing more than a bunch of artificial optimism to me," continued Grason. "I am a realist. I don't think there is anything to be gained from all of this cheerleading." "I'd like to change your mind," said Brown, "but I'm not sure where to start."

Discussion Questions

1. Assess Grason's comments about a can-do attitude. How will her attitude affect those of her direct reports at TAI?
2. Put yourself in Brown's position. How would you explain the importance of a can-do attitude to Grason?

CASE 4.2   How Can I Get a Can-Do Attitude?

"Before I was CEO of this company, I spent 20 years as a safety engineer. It was my job to be pessimistic—to look for anything that might go wrong and assume it would. I did that well, which is why we have always had such an excellent safety record. Now that I'm the CEO, I'm supposed to have a positive, can-do attitude." Mark Bellanger, the new CEO of Bellanger Technologies, Inc. (BTI), was talking to the retired CEO he had replaced less than a year ago—his father, the founder of BTI. "I don't know if I can make the transition, Dad. You were born with a can-do attitude—I wasn't. How am I supposed to go about getting one?"

Discussion Questions

1. Evaluate Bellanger's comment to his dad: "You were born with a can-do attitude—I wasn't." Do you agree or disagree with this statement? Why?
2. Put yourself in the senior Bellanger's place. How would you answer your son's question about developing a can-do attitude?

CASE 4.3 When Things Get Hectic, You Just Make Them Worse

"Mary, you have got to step back and get a grip on yourself. Every time things get hectic around here, you just make matters worse by becoming agitated and yelling orders at employees. And they either get their feelings hurt or become frazzled themselves." Jean Anderson, CEO of Able Publication Services (APS), was talking to Mary McBride, the new head of the company's copyediting department. "But we cannot afford to miss our deadlines," argued McBride. "These new contracts are critical." "I know. The deadlines are very important," agreed Anderson. "That's why you have to learn to settle down. When things get hectic around here, you just make them worse by becoming frantic. When the employees see you frantically dashing about, they get addled and start making mistakes. Then we just end up doing the work over, and we are even later than we would have been in the first place." "Jean, you always stay calm when things get rushed. In fact, back when you had my job, the more hectic the work schedule became, the calmer you got. How do you do that?" asked McBride.

Discussion Questions

1. Why does McBride become so agitated when the work schedule is rushed?
2. Put yourself in Anderson's place. How would you explain to McBride how to develop the ability to remain calm during a storm?

Endnotes

[1] John C. Maxwell, *The 21 Irrefutable Laws of Leadership* (Nashville, TN: Thomas Nelson, 1998), 153.

[2] Steven F. Hayward, *Churchill on Leadership* (Rocklin, CA: Forum, 1998), 115.

[3] Maurice Hankey as quoted in Steven F. Hayward, 115.

[4] John C. Maxwell, 163.

[5] John C. Maxwell, 163.

[6] Steven F. Hayward, 116.

[7] Laura Chamberlain, "The Lost Art of Leadership," www. womentodaymagazine.com/career/thelostartofleadership.html, October 2, 2003.

[8] John C. Maxwell, *Developing the Leader within You* (Nashville, TN: Thomas Nelson, 1993), 105.

[9] Louis E. Boone, *Quotable Business*, 2nd ed. (New York: Random House, 1999), 57.

[10] Steven F. Hayward, 125–126.

[11] Thomas J. Neff and James M. Citrin, *Lessons from the Top* (New York: Currency/Doubleday, 2001), 181–186.

[12] Thomas J. Neff and James M. Citrin, 182.

[13] James M. Kouzes and Barry Z. Posner, *The Leadership Challenge*, 3rd ed. (San Francisco: Jossey-Bass, 2002), 178

[14] Steven F. Hayward, 46.

[15] James M. Kouzes and Barry Z. Posner, 178.

[16] James M. Kouzes and Barry Z. Posner, 179.

[17] James M. Kouzes and Barry Z. Posner, 179.

[18] Alan Axelrod, *Profiles in Leadership* (Upper Saddle River, NJ: Prentice Hall, 2003), 536–542.

[19] Alan Axelrod, 539–540.

[20] "The Berlin Airlift: Operation Plainfare," http://www.britains-smallwars. com/cold-war/berlin-airlift.htm, February 20, 2003.

[21] Alan Axelrod, 298–299.

[22] Alan Axelrod, 299.

[23] Donald T. Phillips, *Lincoln on Leadership* (New York: Warner Books, 1992), 102–103.

[24] Donald T. Phillips, 103.

[25] Booker T. Washington, *Up from Slavery* (New York: Oxford University Press, 1995), 62–101.

[26] William L. Andrews, inside front cover of *Up from Slavery*, by Booker T. Washington.

[27] Booker T. Washington, 63.

[28] Booker T. Washington, 101.

Develop Self-Discipline, Time Management, and Execution Skills

"All great leaders have understood that their number one responsibility was for their own discipline and growth. If they could not lead themselves, they could not lead others."[1]

—John C. Maxwell
Leadership Expert

OBJECTIVES

- Explain the concept of self-discipline.
- Explain the importance of self-discipline to those who would lead.
- Determine your personal self-discipline quotient.
- Explain the relationship of self-discipline and time management.
- Explain the relationship of self-discipline and effective execution.
- Summarize the self-discipline lessons of selected leaders.

John is a mechanical engineer with a responsible position in a growing company. He is good at what he does. As the company grows, John hopes to grow with it. But he has a problem. John is a procrastinator—he puts things off until the last minute and as a result is often late completing assignments, arriving at meetings, and fulfilling other obligations. John is also a poor manager of his time. Consequently, he often wastes the time of others. John knows his poor time management and procrastination might limit his potential for advancement, and he wants to do better, but he just does not seem to be able to break old habits. John's problem is that he lacks a key leadership skill—self-discipline.

Many people decide to diet, but few stick with it. Many people start on an exercise program, but few work out regularly. Many people make New Year's resolutions, but few keep them. In each of these cases, the missing ingredient is self-discipline. Once started on a course of action, whether it be a diet, exercise program, organizational change, or any other endeavor, people need self-discipline to stay the course. Those who fail to follow through on plans, programs, and commitments often rationalize their failure by saying, "I just don't have any self-discipline."

In reality, people who use this excuse have as much self-discipline as anyone else. The difference is that they do not apply it. Self-discipline is not something one simply has or does not have. It is neither a gift nor a genetic characteristic. To say, "I have no self-discipline" is like saying, "I can't play tennis." Like tennis, self-discipline is a skill—something that can be learned, and once learned, must be practiced consistently. Technical professionals who hope to be effective leaders must develop self-discipline. This point cannot be overemphasized. Remember this:

> You must learn to lead yourself first. Only then can you lead others. Self-discipline is self-leadership.

SELF-DISCIPLINE DEFINED

Self-discipline is the ability to consciously take control of your personal choices, decisions, actions, and behavior. Two key aspects of this definition are found in the terms *ability* and *conscious*. The term *ability* is important because it conveys the message that self-discipline—like any other ability—can be learned and then developed through consistent effort over time. The term *conscious* is important because it conveys the message that self-discipline is a choice, rather than a trait one just happens to have or not have.

People who exercise self-discipline are just as susceptible to the typical propensities of human nature as anyone else. The difference between those who do and those who do not exercise self-discipline is choice. Those who do not exercise self-discipline tend to think that making the right choice in a given situation is somehow easier for those who do. It isn't. Self-disciplined people are just as tempted as anyone else to take the easy, comfortable, or expedient way. The difference is that they consciously choose to make the right decision and do the right thing. Self-disciplined people are self-disciplined because they choose to be.

Exercising self-discipline might amount to something so seemingly inconsequential as getting out of bed when you feel like sleeping "*just a little longer.*" It might involve controlling your temper when giving constructive criticism to a poorly performing employee or when debating with a pushy, self-interested colleague. It might mean staying behind to finish the work on an important bid proposal when good friends have invited you out for an evening on the town. It almost always means choosing option B when option A would be easier, more comfortable, or more expedient.

There is an element of self-denial implicit in the application of self-discipline. Often the principal barrier to doing what is right when another option is more appealing is an aversion to self-denial. People who will not deny themselves the luxury of the easy, comfortable, or expedient option have not yet developed self-discipline.

IMPORTANCE OF SELF-DISCIPLINE

Often the key difference between success and failure is self-discipline. The road to success is paved with the unmet potential of talented people who failed because they lacked self-discipline. This is why self-discipline is so important to those who would lead. Self-discipline manifests itself in several different ways, all of them positive. Consider the following benefits of self-discipline:

■ *Time management.* To be an effective leader, technical professionals have to learn to make efficient use of their time. The many tasks of the leader, taken together, are always time-consuming. Consequently, good time management is important to leaders, and good time management requires self-discipline. Self-disciplined leaders make effective use of their time, which will in turn contribute to better performance.

Leadership Tip

"If you could kick the person responsible for most of your troubles, you wouldn't be able to sit down for weeks."[2]

—John C. Maxwell
Leadership Expert

■ *Stewardship.* Effective leaders are good stewards of the financial, technological, and human resources entrusted to them. Overseeing all of these various resources while still tending to the everyday aspects of the job (paperwork, meetings, email, telephone calls, etc.) takes both time and consistent effort. Consequently, self-disciplined leaders, because they manage their time well and because they put forth the necessary effort, are more likely to be good stewards. This is important because good stewardship contributes to better performance.

■ *Execution.* Self-disciplined leaders are more efficient and effective in executing their plans. Even the best strategic or operational plan in the world is just a dream until it is effectively executed. Effective execution requires self-discipline. This is critical because in a competitive environment the difference between winning and losing is often effective execution.

■ *Followership.* Self-disciplined leaders are better able to do all of the things necessary to win and maintain followership (e.g., consistently set a positive example, maintain a can-do attitude, establish and maintain credibility) because they do these things even on days when they don't feel like it. Consequently, self-discipline is fundamental to effective leadership because it helps establish and maintain followership.

The value of self-discipline can be seen in the following scenario involving two technical professionals. John and Mark are department heads for Safe-Tech, Inc. Although they are the same age, graduated from the same university with the same degree, and have the same amount of work experience, their departments perform at radically different levels. John's department typically has the highest performance rating in the company. Mark's department typically has the lowest. In spite of the many similarities between John and Mark, these two technical professionals differ significantly when it comes to self-discipline.

John is an excellent time manager. Whenever he conducts a meeting, there is always an agenda with both a starting and projected ending time. In addition, all participants receive the agenda—along with backup material—at least 24 hours before the meeting convenes. This approach minimizes the amount of wasted time spent in John's meetings. Mark, on the other hand, is known for conducting meetings that are long on time, but short on organization. He does not use an agenda, and he frequently leaves participants idly waiting while he leaves the room to search for backup material he forgot to bring to the meeting.

John is a good steward of the resources for which he is responsible. He sees to it that his direct reports receive the training and mentoring they need to do their jobs and to grow professionally. Because he is a wise and careful budget manager, John is able to keep the technologies in his department up-to-date and operating at maximum efficiency. Mark, on the other hand, releases his direct reports for training only when it is absolutely necessary. When it comes to budget management, Mark's philosophy is "I'm an engineer, not an accountant." As a result, Mark's departmental budget often runs dry before the end of the fiscal year. He does try to keep his department's technologies up-to-date, but because he is a poor budget manager, he seldom has the necessary funds.

John's time management and good stewardship are direct benefits of his self-discipline. Mark's poor time management and weak budget management are the result of a lack of self-discipline. Mark could take the time to develop an agenda and organize backup materials for meetings as John does, but he chooses not to. Mark could monitor his departmental budget daily as John does, but he chooses not to. As a result of his undisciplined behavior, Mark is a much less effective leader than John, and it shows in his department's performance.

ASSESSING YOUR SELF-DISCIPLINE

Are you a self-disciplined person? The best way to answer this question is by conducting a self-assessment. Figure 5.1 is a self-assessment instrument you can use to identify your current level of self-discipline.

The questions in Figure 5.1 represent some minimum-level expectations of leaders. Self-disciplined leaders would be able to answer "Yes" to all 10 questions, giving them a self-discipline quotient of 100. Any question answered "No" indicates a need to improve your self-discipline. The functional word in each question is *consistently*.

A "Yes" answer to a question means that you do what the question asks day after day, week after week, and month after month. This does not mean that you are never late for work or that you never let a meeting run overtime. Even the most self-disciplined person could make no such claim, because nobody has the control over events and circumstances that would be required for a perfect record. After all, something as unpredictable and uncontrollable as a traffic jam can make you late for work. And in meetings, running overtime might be warranted if the topic of discussion is of sufficient importance. What a "Yes" answer to any of the questions in Figure 5.1 does say, however, is that "this is the case except in rare instances." With this clarification, a self-disciplined leader should be able to answer "Yes" to all 10 of the questions.

Leadership Tip

"People count up the faults of those who keep them waiting."

—French Proverb

Self-Assessment
Self-Discipline Quotient

Answer each of the following questions "Yes" or "No." Each Yes response gives you 10 points. Each No response gives you 0 points and indicates a need to improve your self-discipline.

Yes	No	
❑	❑	1. Do you consistently arrive at work on time or early?
❑	❑	2. Do you consistently arrive on time for meetings, appointments, and other similar obligations?
❑	❑	3. Do you consistently submit work assignments on time or early?
❑	❑	4. Do you consistently keep up-to-date in your professional field?
❑	❑	5. Do you consistently keep up with your work-related reading?
❑	❑	6. Do you consistently and promptly return telephone calls?
❑	❑	7. Do you consistently and promptly return email?
❑	❑	8. Do you consistently begin meetings on time?
❑	❑	9. Do you consistently end meetings on time?
❑	❑	10. Do you consistently and promptly follow through on promises?

FIGURE 5.1   What is your self-discipline quotient?

SELF-DISCIPLINE AND TIME MANAGEMENT

Self-disciplined leaders are good time managers. They have to be because there are so many competing demands for their time. Leaders who fail to manage their time well find themselves also falling short on critical leadership tasks as a result. Figure 5.2 lists some of the problems that can result when leaders manage their time poorly. The various problems taken together can lead to even greater problems that rob the organization of its ability to perform at a competitive level. These performance-oriented problems,

**Leadership Problems Caused by
Poor Time Management**

- Wasted time (yours and that of others)
- Stress (on you and others)
- Lost credibility with followers
- Missed appointments
- Poor follow-through on commitments
- Inattention to detail
- Ineffective execution
- Poor stewardship of resources

FIGURE 5.2   Typical problems associated with poor time management.

**Performance Problems Caused by
Poor Time Management**

- Poor quality
- Decreased productivity
- Declining customer satisfaction
- Insufficient attention to continual improvement

FIGURE 5.3   Leaders who cannot manage their time harm their organization's performance.

shown in Figure 5.3, include quality, productivity, customer satisfaction, and continual improvement problems.

The following scenario shows how poor time management can affect these performance factors. Myra X is director of the remodeling and renovations department for a large construction company. She has excellent credentials and a strong background in construction management, but her department never seems to live up to its potential. Earlier in her career, Myra exercised good self-discipline. But in more recent years, she has fallen into some bad habits and let her self-discipline slip almost to the point of nonexistence. In fact, she has become a notoriously poor time manager.

Because she is always running late, Myra never seems to have time to spend with employees or customers. As a result, she typically ignores their complaints, concerns, ideas, and suggestions. Her most frequent response to employees, colleagues, and customers is "I'll get with you later on that. I don't have time right now." Of course, concerns and complaints that are ignored—whether from employees, colleagues, or customers—typically lead to problems that just get worse as time goes by. The longer they are ignored, the bigger the problems get. This happens to Myra all the time.

Myra has also become a procrastinator. She puts off things that need to be done until they cannot be put off any longer. Then she rushes around neglecting every other responsibility in her life until the most pressing obligation of the moment is finally satisfied. As soon as one fire is put out, she begins the cycle all over again: procrastinate-panic-rush ... procrastinate-panic-rush. As a result of Myra's procrastination and poor time management, her direct reports are often stressed out and irritable. The tardiness, absenteeism, and turnover rates in her department are the highest in the company. In addition, Myra's customers are typically frustrated and angry about missed deadlines and rushed, last-minute work that too often is sloppy or half done. Very few customers give Myra any repeat business, and they seldom refer new clients to her.

All of the various problems associated with Myra's poor time management, taken together, have caused some of her key direct reports to become overwhelmed by the constant frustration and to just give up. These key employees no longer help the company compete, because they just don't try any more.

TIME MANAGEMENT PROBLEMS AND WHAT TO DO ABOUT THEM

Most technical professionals have had the experience of coming to work with a good idea of what they hope to accomplish that day only to have most of their time taken up by unplanned activities. The most common causes of time management problems are predictable: unexpected crises, telephone calls, no planning or poor planning, biting off more than you can chew, unscheduled visitors, refusal to delegate, disorganization, lack of technology skills, refusing to say "no," and meetings.

Unexpected Crises

Crises go with the job for most technical professionals. Although it is a fact that the better you plan, the fewer crises you will face, even with good planning, crises will happen. Events over which you have no control can create unexpected circumstances that must be dealt with. Consequently, it is wise to assume there will be crises and act accordingly. The following strategies can minimize the amount of time devoted to putting out fires.

Do not overbook.   If you have ever been to a doctor's office, you are familiar with the problem of overbooking. Typically, by midmorning the doctor's schedule has already gotten backed up. All it takes is one crisis with a patient, and the rest of the day is thrown off schedule. By scheduling loosely, you build in time to deal with the predictable occurrences of crises. The key to loosening your schedule is learning to occasionally say "no."

Do not become an adoption agency for the unrelated problems of others. Leaders are can-do people who, when they see a problem, take the initiative and solve it. Normally this is precisely what a leader should do, but not always. Occasionally an employee will bring you a problem that falls outside your range of responsibility or sphere of influence. Often such problems are personal in nature. In such cases, there is nothing wrong with offering advice or making a helpful referral. In fact, providing this type of help is recommended. But advice and referral are as far as your help should go. Do not become an adoption agency for the personal problems of your direct reports. Problems that relate to your job and areas of responsibility should receive your full attention. Unrelated problems must not be allowed to occupy your time.

Telephone Calls

Unless you manage its use wisely, the telephone can rob you of valuable time. A log of your calls will make the point: For just one week, monitor the amount of time you waste threading your way through the seemingly unending options available on many computerized answering systems. Unless you know the actual extension of the person you are calling, a great deal of time can be wasted listening while a computerized voice gives you a lot of irrelevant options. Cellular telephones have only served to magnify the amount of time in the typical day taken up by telephone calls. Fortunately, cellular telephones can also help you solve this problem. The following strategies will help minimize the amount of time in your day that is tied up by the telephone.

Increase your use of email. One of the best ways to avoid wasting time on hold, playing telephone tag, or listening to an endless array of irrelevant options is to use email wherever practicable. Email is not always a feasible option. However, when it is, you simply click on "Send" and move on to your next task—no pressing 1 for this or 2 for that, no talking to answering machines, and no waiting on hold. In addition, preliminary research shows that people are better about promptly returning email queries than they are about returning telephone calls.

Separate important, routine, and unimportant calls. Time invested in helping secretarial personnel learn to distinguish between important and unimportant telephone calls and between important and routine calls is time well spent. One of the ways to do this is to provide secretarial personnel with a priority list (e.g., put these callers through every time; put these through unless I am really busy). In addition, train secretarial personnel to take comprehensive, detailed telephone messages. This will help you determine which calls should be returned and in what order.

Return telephone calls between meetings and during breaks. Cellular telephones can be obnoxiously intrusive. How many times has a meeting or a conversation in which you were involved been interrupted by the inopportune ringing of a cellular telephone? On the other hand, cellular telephones can help you turn time that might otherwise be wasted into productive time. You can save valuable time by taking telephone messages to meetings with you and using your cellular telephone to return them during breaks and between meetings. You can also use them to return calls from your car while traveling to your next meeting—provided you have the appropriate "hands off" technology or that your car is off the road and parked. Do not be one of those people who drive with their knees while dialing a cellular telephone. Remember, you cannot lead from a hospital bed.

Block out call-return times on your calendar. Telephone tag is one of the most frequent time wasters in the workday. Say, for whatever reason, that email is not appropriate and you really need to talk to someone. You place the call, but the person you need to talk to is not available. You leave a message. This person wants to talk with you, too, so she calls you back, but you are tied up in a meeting. She leaves a message. You call her back, but just miss her. . . . Sound familiar? This frustrating process repeats itself continually every day in businesses everywhere in the world. To minimize the amount of time you waste playing telephone tag, block out times on your calendar for returning calls, and let the times be known to callers who leave messages. An effective approach is to schedule two 30-minute blocks (at least) in each day, one in midmorning and one in midafternoon. Block these times out on your calendar as if they are appointments. Make sure that whoever takes your messages (or your recording device) lets callers know that these are the times during which you typically return calls. In this way if the caller really needs to talk with you, she will make a point of being available during one of these times. In fact, she can make connecting even easier by indicating which time she prefers.

Get to the point. One of the reasons telephone calls are such time robbers is the human propensity for idle chitchat. You can save a surprising amount of time on the telephone by simply getting to the point and by tactfully nudging callers to do the same. There is certainly nothing wrong with a few appropriate comments on the latest ball game, stock prices, or terrorist threats, but the amount of time devoted to unrelated issues should be kept to a minimum. Stay focused, stay on task, and help callers learn to follow suit.

Poor Planning

It has already been mentioned that crises are a common time waster for leaders. There are crises that simply cannot be avoided, and there are crises that result from poor planning. A good rule of thumb to remember is this:

The more effort you put into planning, the less effort you will waste dealing with crises. The following strategies can help improve your planning.

End each day by planning for the next.   Devoting just 10 to 15 minutes at the end of each day to reviewing and planning for the next day can save valuable time. One of the best ways to minimize the occurrence of crises is to already know when you walk in the door of your office each morning what is on your plate for the day. Three of the most persistent causes of crises are forgotten meetings, overlooked obligations, and missing documentation. You can eliminate these causes altogether by ending each workday by making a quick review of the next day's meetings and obligations, getting out any paperwork or other documentation that will be needed and familiarizing yourself with it, and creating a "to do tomorrow" list with the items listed in priority order. With just these few tasks accomplished, you will be able to begin the next day ready for what awaits you.

Consider that most tasks take longer to complete than you think they will. This is a good rule of thumb to follow. No matter what it is you have planned to do, experience shows that it will probably take longer than you think it will. Consequently, it is wise to build a little extra time into your schedule. For example, if you think an appointment will take 30 minutes, allow 45 minutes, then try to finish in 30. In this way you have the extra time if it is needed without rushing through the appointment. If the appointment concludes on time, you can always put the extra time gained to good use returning telephone calls or getting a head start on other obligations.

Trying To Do Too Much

Leaders are people who take the initiative and seek responsibility. This is one of the reasons they stand out from the crowd and become leaders. It is also the reason for one of those ironic situations in which the good news and the bad news are the same. The good news is that leaders take the initiative; the bad news is that leaders take the initiative. When you take the initiative and seek responsibility, it is easy to take on too much and get out of balance. When this happens, the following strategies can help you free up some valuable time.

Write down all current and pending tasks, projects, and obligations. Make a list of all the commitments on your plate. Then prioritize them. For each entry on your list, ask the following question: What will happen if I don't do this? This will usually give you a good start on paring down the list. Once you have eliminated obligations you don't really need to do, prioritize the rest.

Examine all outside activities.   Technical professionals, especially those who are leaders and those who aspire to be, are typically active in outside organizations such as civic clubs, chambers of commerce, economic development councils, and professional organizations in their fields. Participation in these and other outside activities is an excellent way to grow as a leader and to benefit your organization. However, it is easy to fall into the trap of taking on too many outside activities and responsibilities. Balance is the key. For technical professionals who aspire to be effective leaders, outside activities are like food. Certain amounts of the right types are essential, but too much—even of the right types—can be bad for you.

Unscheduled Visitors

One way to partially regulate your time is to encourage people who need to see you to make an appointment. People who just drop in unannounced can rob you of valuable time. The following strategies can help you minimize the amount of time lost in your day to drop-in visitors.

Do not allow drop-in visitors during peak times.   Some days are busier than others, and some times of the day are busier than other times. During these peak times, it is best to tell drop-in visitors to come back at another time when you can give them your undivided attention—unless they are bringing you critical information or informing you of an emergency.

Train secretarial personnel to rescue you.   You can minimize the intrusions of drop-in visitors by working out an arrangement with secretarial personnel to rescue you after a set amount of time (e.g., five minutes). It works like this. Whenever a drop-in visitor has been in your office for five minutes or so, the secretary buzzes you or looks in and says "It's time to place that important call," or "It's time for your next meeting." This will tactfully get the drop-in visitor on his way.

Remain standing.   One way to convey the message that you are busy without having to actually say it is to remain standing when an unannounced visitor walks into your office. Once a visitor sits down and gets comfortable, it can be much more difficult to uproot him. By continuing to stand, you convey the message that, "I can give you a few minutes, but only a few."

Poor Delegation

Poor delegation is one of the easiest time-wasting traps to fall into. Often, technical professionals who become leaders in their fields find it difficult to let go of work they are accustomed to doing themselves. In addition, some suffer from the "nobody can do it right but me" syndrome. These two phenomena can result in poor delegation, a major time waster. Tasks that do not require your level of expertise should be delegated. If subordinates cannot perform the tasks satisfactorily, you have either a training or a human resources problem, neither of which should be solved by refusing to delegate work.

Personal Disorganization

You can waste a lot of time rummaging through disorganized stacks of paperwork looking for the folder, form, or document needed. The author once worked with an individual who had the unfortunate habit of never putting files, documents, or forms in the same place twice when he was done with them. Wherever the engineer happened to be when he finished with a file is where he would put it down. As a result, this otherwise talented engineer wasted an inordinate amount of time looking for "missing" paperwork. The following strategies can help minimize the amount of time lost in your day as a result of personal disorganization.

Clean off your desk.   This strategy sounds so simple that one might be tempted to discount it. But before doing so, look at your desk. Check your in-basket and your pending basket. Is there paperwork that is no longer relevant or should be filed? Go through everything on your desk and get rid of anything that is no longer pertinent. When trying to get organized, your best friend can be a large trash can.

Restack your work in priority order.   Go through your in-box, pending file, and to-do stack and organize all work in order of priority. Work is often stacked in the order it comes in, especially when you are in a hurry and do not have time to organize it. Because this can happen frequently, it is a good idea to occasionally take a few minutes to go through your work stacks and reorganize everything in them by priority.

Screen incoming work.   Screen incoming work before putting it in your work stack or pending file. If it can be gotten rid of, do it. Paperwork that is important should be placed in the stack according to its priority. Just placing work on top of the stack as it comes in can create two problems. First, unimportant and routine paperwork often will be placed on top of high-priority work, which can cause important work to go unnoticed. Second, unprioritized paperwork forces you to waste time flipping through irrelevant material looking for what is important.

Make use of categorized work files.   Teach clerical personnel to organize your paperwork by category. This means having a "read" folder for paperwork that should be read but requires no writing or other action. Have a "correspondence" folder for nonelectronic correspondence you need to answer or initiate. Have a "signature" folder for paperwork that requires your signature (correspondence, requisitions, etc.). Organizing work in this way allows you to get right at what needs to be done without having to waste time sorting through stacks of paperwork.

Inefficient Use of Technology

Even technical professionals are sometimes guilty of inefficient use of technology. Time-saving technologies save time only if you know how to use all of their various features and use them well. The following strategies can help improve how efficiently and effectively you use time-saving technologies:

- Make sure you can use all of the various functions on your regular telephone and your cellular telephone (e.g., messaging, call waiting, park, camp, automatic redial, forward, save, repeat).
- Learn to dictate correspondence using a microcassette recorder.
- Make sure you can use all of the icons and other features on your computer.
- Make sure you can operate your fax machine.
- Make sure you can use all of the time-saving functions on your copy machine (sorting, collating, etc.).
- Equip your car as a second office.
- Learn to use all of the features of your electronic calendar.

Unnecessary and Inefficient Meetings

In spite of their value in bringing people responsible for different functions together to convey information, brainstorm, plan, and discuss issues, meetings can be one of the leader's biggest time wasters. There are two types of meetings that can rob you of an inordinate amount of time: unnecessary meetings and meetings that are too long. The key to getting the most value out of meetings is to meet only when necessary and keep necessary meetings as short as possible. Following are some strategies that will help in this regard.

Understand what causes the wasted time associated with meetings. Informal research conducted by the author suggests that at least 50 percent of the time spent in meetings is wasted. There are many reasons for this, the most prominent of which are as follows: poor preparation, the human need for social interaction, idle chit-chat, would-be comedians, interruptions, getting sidetracked on unrelated issues (loss of focus), no agenda, and no prior distribution of backup materials. In addition to these time wasters, there is also the "comfort factor." Coffee, goodies, social interaction, and proximity to the boss can create such a desirable environment that people simply do not want meetings to end.

Examine all regularly scheduled meetings carefully. Most organizations have weekly, biweekly, and monthly meetings of various groups. When these meetings were established, they had a definite purpose, but over time the purpose may have become blurred and the meetings continue only out of habit. If you call or attend regularly scheduled meetings, ask the following questions about them: (1) Is the meeting really necessary? (2) What is the purpose of the meeting? (3) Could the meetings be scheduled less frequently? and (4) Could the purpose of the meeting be satisfied some other way (email updates, written reports, etc.)?

Hold impromptu meetings standing up.   Meetings that should last no more than 10 minutes can be kept on schedule by holding them standing up. These are typically impromptu meetings without an agenda called to quickly convey information to a select group. Also, try to hold them in your office rather than in a conference room. It will be hard to keep participants from pulling up chairs and settling in if the meeting is held in a conference room.

Before meetings, complete the necessary preparations.   Have an agenda that contains the following information: purpose of the meeting, starting and ending time, list of agenda items with a responsible person for each, and a projected amount of time to be devoted to each agenda item. Set a deadline for submitting agenda items and stick to it. Require all backup material to be provided at the same time as the corresponding agenda items. Distribute the agenda, backup material, and the minutes of the last meeting at least two hours before the meeting. If you distribute meeting materials too far in advance, participants will simply put them aside and forget about them. In addition, you increase the likelihood of cutting off the submittal of agenda items too soon. If you wait until the meeting to distribute materials, you will waste time handing them out and waiting while participants read them. Require all participants to read the agenda and backup material before the meeting; this is why it is distributed beforehand in the first place.

During the meeting, stay focused and stick to the agenda.   Begin meetings on time. Waiting for stragglers only encourages tardiness. If participants know you are going to start on time, they will discipline themselves to arrive on time. Have someone take minutes. In the minutes, all action and follow-up items should be typed in boldface so that they stand out from the routine material. Make the minutes of the last meeting the first item on the agenda. In this way the first action taken in a meeting is following up on assignments and commitments made during the last meeting. Stay focused. Keep participants on the agenda and on task. The last agenda item should always be either "new business" or "comments from the floor." Such an agenda item gives participants an opportunity to bring up issues that are not on the agenda without getting the meeting sidetracked before agenda items have been disposed of. Leaders should make sure that only critical issues and bona fide emergency cases are brought up as new business. Otherwise, participants will begin to bring up all of their issues in new business rather than devoting the preparation time necessary to get them on the agenda. Require participants to turn off their cellular telephones. Interruptions from cellular telephones have become a major distraction and time waster in meetings.

After meetings, follow up quickly.   Have the minutes typed and distributed right away—ideally on the same day as the meeting. Email distribution can help in this regard. Allow an appropriate amount of time for participants to act, then follow up on action items from the minutes. If you call meetings, never wait until the next meeting to ask about progress made toward completing the action items from the previous meeting.

EFFECTIVE EXECUTION AND SELF-DISCIPLINE

The literature on leadership is replete with articles on strategic planning, operational planning, and other creative activities. In fact, the creative process is so valued in business and leadership circles that the phrase "thinking outside the box" has achieved the status of mantra—and rightfully so. Creativity is critical to good planning, and planning is critical to global competitiveness. Planning, as a leadership activity, has star quality; it is clean, creative, visionary, and fun. But executing a plan is a different story. Execution is a roll-up-your-sleeves and get-your-hands-dirty type of activity. When developing a plan you can *dream*, but when executing a plan you must *do*.

This is one of those situations in which the saying "The devil is in the details" applies. It is during the execution of a plan that unforeseen difficulties arise to block progress, that seemingly good ideas arrived at with the best of intentions lead to unintended consequences, and that overlooked details spawn intransigent problems. Following are some strategies that can help ensure the effective execution of any plan at any level of an organization. The self-discipline to stay focused is important to all of these strategies.

Strategies for Effective Execution

As a leadership function, effective execution has only recently begun to get the attention it deserves. For years the business literature devoted to execution was sparse at best. Fortunately, the void is being rapidly filled. For example, Larry Bossidy and Ram Charan in their book *Execution: The Discipline of Getting Things Done* list the following strategic leadership behaviors that promote effective execution:[3]

- Know your people and your business.
- Insist on realism.
- Set clear goals and priorities.
- Follow through.
- Reward doers.
- Expand people's capabilities.
- Know yourself.

Leadership Profile   Jim Brodhead Executes at FPL Group[4]

Self-discipline manifested at the organizational level contributes greatly to effective execution, and effective execution is critical to success in a globally competitive environment. Most companies can develop a good strategic plan, but effectively executing the plan is a much different proposition. To do what is necessary to effectively execute its plans, a company must exercise self-discipline at the organizational level.

Bringing the discipline necessary for effective execution to his organization was one of Jim Brodhead's major accomplishments as CEO of utility giant Florida Power & Light (FP&L). When Brodhead took over as CEO, the company was still called FP&L (it became FPL Group under his leadership). He inherited a company that was profitable but complacent. According to Brodhead, FP&L was "a slow-moving, inflexible, bureaucratic company."[5] Success can breed complacency in organizations, and in a globally competitive environment complacent companies can become losers overnight. Brodhead understood that if his company was not moving forward, regardless of current profit levels, it was in reality going backward. He saw problems looming on the horizon if FP&L continued on the same course. But ironically, the company's profitability had instilled an attitude of "if it's not broke, don't fix it" in key personnel. Brodhead's first challenge was to convince these key personnel that the company had a problem.

Convincing them took some effort, but once Brodhead had, he involved them and a broad base of company personnel in developing a vision. With the vision in place, Brodhead led the company through the development of a plan for pursuing the vision. The plan focused on four key elements: (1) improving quality, (2) becoming customer oriented, (3) getting faster and more flexible, and (4) enhancing the company's cost position. With a good plan in place, Brodhead shifted his attention to execution. This is where his leadership really came into play.

At this point, Brodhead was like the quarterback of a football team going into a championship game: He has a sound game plan, but as he stares down the field at the beckoning end zone, he knows that the difference between winning and losing will be execution. Under Brodhead's leadership, FPL Group did indeed execute. The company improved its cost position, enhanced quality, became customer oriented, and became faster and more flexible. In the process the old-fashioned, inflexible, bureaucratic company was transformed into a world-class competitor.

The strategies Brodhead used to ensure effective execution were as follows: (1) involve a broad base of personnel in the planning process to ensure buy-in; (2) communicate, communicate, communicate; (3) empower employees to act; (4) train, train, train (give employees the knowledge and skills needed to execute the plan); (5) establish performance incentives tied directly to the plan and reward performance; (6) establish effective measurements and monitor progress; and (7) celebrate progress and successes. Using these strategies, Brodhead ensured effective execution of the company's four-point plan and transformed a parochial power company into a diversified global competitor.

Of these seven strategies, two in particular require self-discipline. To know your people and your business, you have to be disciplined enough to put in the time and effort it takes on a continual basis. If you are not a good time manager, you will never get around to doing what it takes to know your business and your people. Follow-through also requires self-discipline. Often the difference between those who just make promises and those who keep them is self-discipline.

Another business researcher who has studied the issue of execution is Melissa Raffoni. According to Raffoni, "The execution phase forces you to translate your broad-brush conceptual understanding of your company's strategy into an intimate familiarity with how it will happen: who will take on which tasks in what sequence, how long these tasks will take, how much they'll cost, and how they'll affect subsequent activities."[6] Raffoni recommends the following strategies for promoting effective execution:[7]

Maintain your focus.   Executing a plan is like hitting a golf ball. To do it well, you have to stay focused. Staying focused requires self-discipline. In addition to applying self-discipline, it is also important to be realistic, keep things simple, and ensure clarity. These are all focus-oriented elements. This means that, when planning, an organization should set realistic goals in the first place. It does not mean the organization should avoid "stretch" goals. Stretch goals, properly developed and thought out, are realistic. Actions to be taken and tasks to be performed should be kept as simple as possible; complexity is the enemy of execution. The final ingredient in the focus equation is clarity. This means that every person who has even the smallest role to play in executing the plan must fully understand his assignments, duties, responsibilities, and deadlines. Confusion is another enemy of execution. People who do not understand their roles cannot be expected to carry them out effectively. For this reason, the need to communicate effectively is implicit in the concept of clarity.

Develop tracking systems.   When you are executing a plan, there should be no guessing or ambiguity as to progress. Leaders need to know how the implementation is going and where problems exist, and they need to know in a timely manner. For this reason, it is important to establish tracking systems that provide simple, clear-cut answers to the question "How are we doing?"

Establish regular, formal reviews.   When you attempt to execute a plan, it is important to assign selected responsibilities to key people. One person may be assigned more than one responsibility, but no responsibility should be assigned to more than one person. This approach establishes a single point of contact for accountability. Once the responsibilities have been assigned, these key personnel should meet periodically to report on progress in their areas of responsibility. This is the best way to ensure an appropri-

---

Execution Strategies

- Involve those who must execute the plan in developing it.
- Communicate clearly and often with those who must execute the plan.
- Empower employees to take action.
- Equip employees to carry out their responsibilities (e.g., train, mentor, and provide the necessary resources).
- Establish an effective monitoring system and insist on realism.
- Hold periodic review meetings.
- Reward and recognize performance and progress.
- Celebrate successes.

FIGURE 5.4   Executing a plan requires organizational self-discipline.

ate level of accountability and to identify roadblocks that are hindering progress.

It should be apparent from the execution strategies presented here that different experts recommend different (although similar) strategies. Figure 5.4 lists execution strategies the author recommends. These eight strategies can be used by any leader in any kind of technology company to help ensure effective execution when implementing a plan.

LESSONS ON SELF-DISCIPLINE FROM SELECTED LEADERS

Following are excerpts from the lives of several leaders that exemplify some of the self-discipline principles set forth in this chapter. The leaders selected for inclusion are Franklin Delano Roosevelt, John F. Kennedy, and Don Fites.

Franklin Delano Roosevelt[8]

"The four-term 32nd president of the United States, Franklin Delano Roosevelt, led the nation through two of its most challenging crises: the Great Depression and World War II."[9] Only Abraham Lincoln can be said to have served as president of the United States at a time as rife with crisis as the years when Roosevelt served, and no president can claim to have served under more difficult personal circumstances than Roosevelt, who, throughout his presidency, led from the seat of a wheelchair—his legs crippled by the debilitating scourge of polio. The self-discipline lesson from FDR is as follows:

> Through self-discipline a leader can overcome even the most-challenging obstacles that stand between him and success.

That so little was said or written about FDR's condition at a time when society still tended to confuse physical challenges such as polio with physical weakness is a fitting testament not just to FDR's courage and determination but also to his incredible self-discipline. So self-disciplined was FDR in projecting an image of strength and vitality—an essential undertaking for the leader of the free world during World War II—that many Americans were unaware of his painful struggle with polio. Many Americans who saw their president stand at a lectern flashing his famous jaunty smile—his legs strapped in heavy metal braces discreetly covered by his trousers—were unaware that FDR was enduring intense pain.

During his four terms as president, FDR traveled the globe to meet with world leaders and military commanders. This was during a time when there were no special accommodations for people in wheelchairs. FDR had to be especially self-disciplined to overcome all of the day-to-day challenges to his mobility. In fact, once while at sea during World War II, it was necessary for the navy to rig a rope-and-pulley system to convey FDR from one ship to another. As both ships slowed in the choppy seas, the president dangled in mid-air supported by nothing more than thin rope.

When students of leadership discuss FDR, the conversation typically revolves around the calm reassurance he gave Americans as they endured the economic struggles of the Great Depression or his uplifting determination after the devastation of Pearl Harbor. But one should never end a discussion of FDR's leadership without acknowledging the incredible self-discipline displayed by this courageous president.

John F. Kennedy[10]

When he was assassinated in Dallas on November 22, 1963, John Kennedy became the youngest American president to die in office. He was already the youngest person ever elected president (at the time). In spite of his poor health and debilitating injuries, Kennedy was able to create an image of youthful vigor and vitality that struck a chord with many Americans. Many came to see Kennedy as the president who passed the torch of leadership to a new generation of Americans. But it was more than image that connected Kennedy with the American public. He was attractive, bright, articulate, and visionary. He had that all important ability of great leaders—the ability to inspire people to action.[11]

Youthfulness, vigor, and vitality were hallmarks of John Kennedy's presidency. Ironically, what many Americans did not know was that their seemingly fit young president suffered from chronic and occasionally in-capacitating back pain. Some days the pain he suffered was so intense he could not walk. Kennedy's back problems began when he was injured while playing football at Harvard. The injury was exacerbated when the PT boat he commanded in World War II was rammed by a Japanese de-stroyer. In 1955 Kennedy had to endure two major back operations. From that point on for the rest of his life, Kennedy would struggle daily—often requiring a back brace and sometimes crutches— to keep up the busy schedule required of a U.S. senator and, later, a president of the United States. So self-disciplined was the young president in hiding his pain that few even knew about it. The self-discipline lesson from John F. Kennedy is as follows:

> With self-discipline a leader can overcome even debilitating health
> problems and do what is necessary to inspire his followers to excel.

Much was made in the media of the rocking chair President Kennedy sat in during conferences in the Oval Office. It was portrayed as a quaint and comfortable symbol of this young president's informality, but in real-ity it was the only chair he could sit in that did not aggravate his back prob-lems. There are photographs of Kennedy standing next to a table in the Oval Office and leaning on it with both hands as if studying a map or some other document. In reality, leaning on a table or desk like this was one of the few ways the president could relieve his back pain.

John F. Kennedy's accomplishments as a leader have been well docu-mented. He was a PT boat commander in World War II who was decorated for courage under fire; a U.S. senator; and, up until that time, the youngest person ever elected president of the United States. His entire career was spent in the bright glare of the media. He was under constant public scrutiny. As president, he had to keep an incredibly busy schedule, and he had to deal with the most pressure-packed, nail-biting incident in the his-tory of the cold war—the Cuban missile crisis, an incident that brought the world to the brink of a nuclear holocaust.

That Kennedy could achieve all he did while enduring chronic, painful back problems and a variety of other debilitating health problems is com-mendable in and of itself. But to have done so while projecting a public im-age of vigor and vitality is a testament to his self-discipline and strength of will. Kennedy's is an example that might well be emulated by all technical professionals who would be leaders.

Don Fites[12]

Since the end of World War II, Japanese manufacturers of earth-moving equipment have been just as successful as Japanese automobile makers in selling their products globally. A market that was once dominated by American companies now sends a major share of its business to Japan. Business analysts had begun to think that Japanese manufacturers were going to send American companies such as Caterpillar into bankruptcy. But, thanks to Donald V. Fites, the analysts were proven wrong. Under the leadership of this dynamic CEO, Caterpillar not only survived the Japanese onslaught, it increased its market share.[13]

Beginning there in 1956, Fites spent his entire career with Caterpillar, Inc. When he became CEO in 1990, his highest priority was saving his company from the Japanese onslaught, which had already taken huge bites out of the markets of the American automobile and electronics industries. Having worked in various positions with Caterpillar all over the globe, Fites knew his company and its competition inside out. The execution lesson from Don Fites is as follows:

> Having a good plan is important, but effectively executing that plan is even more important.

Fites set a goal for his company that was simple but challenging—outperform Japanese competitors in the global marketplace. Then he led the company in developing a plan to achieve his goal. The plan included a complete corporate restructuring, aggressive cost cutting, and comprehensive cost improvements. The plan was sound, but Fites knew that the key to success was effective execution. Consequently, he didn't just sit back and expect the plan to take care of itself. Rather, to ensure effective implementation, Fites took several steps he saw as necessary to pave the way for those who would have to execute the plan.

Having served in a variety of Caterpillar locations around the world, Fites knew that the decisions necessary to execute his plan could not be effectively made at a central corporate headquarters. He needed to drive decision making down as close as possible to the problems being dealt with. He accomplished this by decentralizing Caterpillar. The decentralization had the effect of driving decision making, responsibility, and accountability down through the ranks. Fites also empowered his personnel to take the actions necessary to execute their portions of the plan, and he ensured that they would take full advantage of their newly acquired empowerment by tying compensation to performance.

According to Fites, Caterpillar was "a company essentially run top down, and my philosophy is to push decision making as far down as you can, to really make people feel accountable and responsible for their decisions, and then reward them when they do well."[14] Fites put a stop to orders from "on high" being passed down the line to employees and began giving employees more control over how well they and Caterpillar performed.

It takes determined self-discipline to effectively lead a diversified company such as Caterpillar. As CEO of this giant earthmover company, Don Fites had the necessary self-discipline, but he also knew how to structure his company in a way that both required and rewarded self-discipline at all levels. Because of Caterpillar's corporate self-discipline under Fites's leadership, the company not only survived the Japanese onslaught but actually thrived during it, increasing its global share of the earthmover market.

Summary

1. Self-discipline is the ability to consciously take control of your personal choices, decisions, actions, and behavior. In this definition the term *ability* conveys the message that self-discipline can be learned and developed through consistent effort over time. *Consciously* is important because it conveys the message that self-discipline is a choice, rather than a trait that one just happens to have or not have.

2. The benefits to leaders of self-discipline are as follows: time management, stewardship, execution, and followership.

3. Self-disciplined leaders are good time managers. They have to be. *Time robbers* that leaders must learn to deal with include unexpected crises, telephone calls, poor planning, trying to do too much, unscheduled visitors, poor delegation, personal disorganization, inefficient use of technology, and unnecessary and inefficient meetings.

4. Effectively executing plans requires not only personal self-discipline for leaders, but self-discipline applied at the corporate or organizational level. Strategies for effective execution include the following: know your people and your organization, insist on realism, set clear goals and priorities, follow through, reward doers, expand people's capabilities, know yourself, maintain your focus, develop tracking systems, establish regular formal reviews, communicate, and train.

5. The life and career of Franklin Delano Roosevelt showed that through self-discipline a leader can overcome even the most challenging obstacles that stand between him and success.

6. The life and career of John F. Kennedy showed that with self-discipline a leader can overcome even the most debilitating health problems and do what is necessary to inspire his followers to excel.

7. The career of Don Fites at Caterpillar, Inc., showed that having a good plan is important for a leader, but executing that plan is even more important.

Key Terms and Concepts

Ability   Conscious
Choice   Develop tracking systems

Establish regular formal reviews

Execution

Followership

Inefficient use of technology

Maintain focus

Personal disorganization

Poor delegation

Poor planning

Self-discipline

Stewardship

Telephone calls

Time management

Trying to do too much

Unexpected crises

Unnecessary and inefficient meetings

Unscheduled visitors

Review Questions

1. Define the concept of self-discipline.
2. Explain why self-discipline is so important to technical professionals who want to be leaders.
3. What is your self-discipline quotient?
4. Explain the relationship of self-discipline and time management.
5. Explain how to overcome the following factors that can rob leaders of valuable time:
   a. Unexpected crises
   b. Telephone calls
   c. Poor planning
   d. Trying to do too much
   e. Unscheduled visitors
   f. Poor delegation
   g. Personal disorganization
   h. Inefficient use of technology
   i. Unnecessary and inefficient meetings
6. List and explain at least eight strategies for effective execution.

LEADERSHIP SIMULATION CASES

The following simulations are provided to generate additional thought and discussion about the principles of leadership explained in this chapter. Readers are encouraged to consider how the situations presented in these cases might apply to them and to discuss the cases with other leaders and leadership candidates.

CASE 5.1 I Just Don't Seem to Have Any Self-Discipline

Meg Critting is not doing as well as her CEO had hoped she would do when she promoted Meg to her current position—vice president for technology development. Critting is an electrical engineer with impeccable credentials and more than 10 years of experience in technology development for Integrated Technologies, Inc. (ITI). When her new position became available, Critting was the obvious choice. In fact, the company did not even advertise the position. Instead, Critting was promoted from within on the basis of her past performance with the company.

But now that she is in an executive leadership position with ITI, some personal characteristics that were not previously apparent are beginning to show. Critting does not seem to manage her time well, nor is she a good steward of the resources for which she is responsible. As a result, the employees in her division have begun to grumble and, worse yet, the performance numbers for Critting's division have continued to slip—even after she has been in the position long enough to have turned them around.

When ITI's CEO discussed the steadily deteriorating situation with Critting, the vice president for technology development could not explain why things were not going well. Finally, she told her boss, "Maybe I'm just not cut out to be an executive-level leader. I don't think I have the self-discipline it takes."

Discussion Questions

1. If you were the CEO of ITI, how would you respond to Meg Critting's assertion that she does not have the self-discipline necessary to be an executive-level leader?
2. Do you think an individual such as Meg Critting can ever become an effective leader?

CASE 5.2 No Time-Management Skills

John Nevering looked at his calendar and the stacks of paperwork awaiting his attention and groaned. Ever since his promotion, Nevering had been buried in work. He came in early and stayed late. He worked on weekends so frequently that his home life had become non-existent. Nevering could not remember the last time he had exercised.

Nevering tried to keep an appointment schedule, but lately his planned schedule didn't seem to matter. Unplanned events, unexpected problems, and unanticipated crises kept cropping up so frequently that he seldom got to his schedule of planned activities. In desperation, Nevering asked a good friend of his to join him for lunch to discuss the situation. "Mike, I am buried. No matter how much time I put in, I can't seem to get caught up," said Nevering. "You have the same job I have, but you don't have to work late very often and I seldom see you working on weekends. And believe me, I would know. I'm here every weekend. How do you do it?" "John, it sounds as if you have a time management problem," said Mike Weldon. "You're right about that," answered Nevering. "What I need to know is what to do about it."

Discussion Questions

1. Have you ever worked with someone who is always behind in their work, late for meetings, or who always seems to have to work late and on weekends? Did this person's poor time management cause problems for anyone else at work?
2. If you were Mike Weldon, how would you answer John Nevering's last question?

CASE 5.3 Good Planning, Poor Execution

Gerald Renfroe had been hired as CEO of Mil-Tech Corporation (MTC) to turn the company around. MTC is a Department of Defense contractor that specializes in the integration of off-the-shelf technologies in military systems and products. The company was founded as a minority firm under Section "8A" of government statutes. Such firms are able to secure government contracts without going through the competitive bid process for a specified period of time. The "8A" program has been very successful at helping minority owned companies get their foot in the door and stay in business long enough to become competitive and eventually stand on their own.

MTC *graduated* from the 8A program 6 years ago and did well until recently. However, over the last 18 months so many key personnel have been recruited away from the company by competitors that MTC has begun to flounder. This is why Gerald Renfroe was hired. The company's board brought Renfroe in for the express purpose of re-engineering the company and making it globally competitive.

Within just weeks of his arrival on the scene, Renfroe knew what needed to be done. He pulled together a re-engineering team consisting of his top management personnel. Within just weeks of the team's first meeting, Renfroe had a sound plan for getting his company back on track. The plan had four major components: cut costs across the board, improve quality continually, establish a customer focus, and branch out into new non-military markets.

All involved were pleased with the plan. Every member of Renfroe's board of directors agreed that the plan was sound. This is why one year later nobody can understand why the plan is not working. When the turnaround plan was completed, Renfroe had given a copy to each of MTC's vicepresidents with a charge to "get started." The plan contained neither specific assignments nor timetables. Nor were there any provisions for monitoring and measuring progress built into the plan. Renfroe expected his vice-presidents to handle the details.

Discussion Questions

1. Analyze this case. Why do you think MTC's turnaround plan is not being executed?
2. If you were Gerald Renfroe, how would you go about ensuring effective execution of the turnaround plan?

Endnotes

[1] John C. Maxwell, *Developing the Leader within You* (Nashville, TN: Thomas Nelson, 1993), 161–162.

[2] John C. Maxwell, 163.

[3] Larry Bossidy and Ram Charan, *Execution: The Discipline of Getting Things Done* (New York: Crown, 2002), 3–12.

[4] Thomas J. Neff and James M. Citrin, *Lessons from the Top* (New York: Currency/Doubleday, 1999), 67–72.

[5] Thomas J. Neff and James M. Citrin, 68.

[6] Melissa Raffoni, "Three Keys to Effective Execution," in *Harvard Management Update* 8, no. 2 (February 2003): 2.

[7] Melissa Raffoni, 2–4.

[8] Alan Axelrod, *Profiles in Leadership* (Upper Saddle River, NJ: Prentice Hall, 2003), 444–453.

[9] Alan Axelrod, 444.

[10] Alan Axelrod, 298–303.

[11] Alan Axelrod, 298.

[12] Thomas J. Neff and James M. Citrin, 123–128.

[13] Thomas J. Neff and James M. Citrin, 123.

[14] Thomas J. Neff and James M. Citrin, 125.

Be a Creative Problem Solver and Decision Maker

"It is when solving problems that a leader's ability to innovate is most apparent. It is when making decisions that the leader's accountability is most apparent."

—The Author

OBJECTIVES

- Explain how to solve and prevent problems.
- Explain the concept of decision making for effective leadership.
- Describe the decision-making process.
- Explain the concept of employee involvement in decision making.
- Explain the concept of creativity in decision making.
- Summarize the problem-solving and decision-making lessons from selected leaders.

Problem solving and decision making are fundamental to effective leadership. The workplace will never be completely problem free, but good decisions decrease the number of problems that occur. This chapter will help technical professionals (1) learn how to solve problems effectively, positively, and in ways that don't create additional problems; (2) become better decision makers; and (3) learn to make decisions and handle problems in ways that promote effective leadership.

SOLVING AND PREVENTING PROBLEMS

Even the best-led organizations have problems. A problem is any situation in which what exists does not match what is desired; or, put another way, the discrepancy between the current and the desired state of affairs. The greater the disparity between the two, the greater the problem. Problem solving is not just "putting out fires" as they occur. Rather, it is one more way to make continual improvements in products or services.

The Perry Johnson Method

Perry Johnson, Inc., of Southfield, Michigan, recommends an approach to problem solving that works well because of its three main characteristics:[1]

1. It promotes teamwork in problem solving.
2. It leads to continual improvement rather than just "putting out fires."
3. It approaches problems as normal by-products of changes.

The Perry Johnson method for problem solving is as follows:

1. *Establish a problem-solving team.* The reason for using a team in solving problems is the same as that for using a team in any undertaking: no single person knows as much as a team. Team members have their own individual experiences, unique abilities, and particular ways of looking at things. Consequently, the collective efforts of a team are typically more effective than the sole efforts of one person.

A problem-solving team can be a subset of one functional unit, or it can have members from two or more different departments. It can be convened solely for problem solving, or it can have other duties. Decisions about how to configure the team should be based on the need, size, and circumstances of the organization.

2. *Brainstorm the problem list.* It is important to get out in front of problems and deal with them systematically. For example, the military doesn't just sit back and wait for the next trouble spot in the world to boil over. Rather, potential trouble spots are identified and entered onto a problem list. The potential problems are then prioritized, and plans are developed for handling them. The same approach can be used in any

organization. The problem-solving team should brainstorm about problems that might occur and create a master list.

3. *Narrow the problem list.* The first draft of the problem list should be narrowed down to the entries that are really problems. To accomplish this, evaluate each entry on the list by means of three criteria:

- There is a standard to which the entry can be compared.
- Actual performance varies from the standard in an undesirable way.
- The variance is supported by facts.

Any entry that does not meet all three criteria should be dropped from the list.

4. *Create problem definitions.* All problems remaining on the list should be clearly defined. A problem definition has two parts: a description of the circumstance and a description of the variance. Figure 6.1 shows sample problem definitions that have the desirable characteristics of being thorough, brief, and precise.

- **Job 21A is over budget by 22 percent.**
  Circumstance:
  Job 21A is over budget.
  Variance:
  By 22 percent

- **Reject rate of the machining department is too high by 12 percent.**
  Circumstance:
  Reject rate of the machining department is too high.
  Variance:
  By 12 percent

FIGURE 6.1 Sample problem definitions.

A teamwork tool that can be helpful in clarifying the definition of a problem is the "why-why" method, which involves asking "why" until the team runs out of answers. Figure 6.2 illustrates how why-why works. Going through this process can be an effective way to determine the agent producing the symptoms.

5. *Prioritize and select problems.* With all problems on the list defined, the team can prioritize them and decide which one to pursue first, second, and so on. Perry Johnson recommends using a problem priority matrix. The matrix is created as follows:

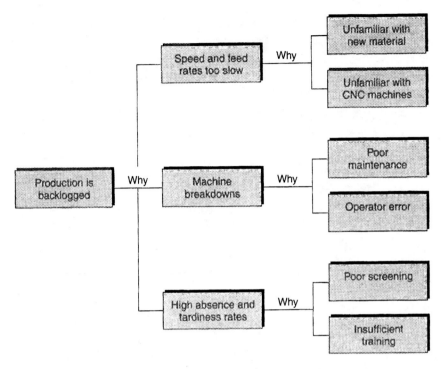

FIGURE 6.2   The why-why method of problem definition.

- Divide the problem-solving team into two groups and put them in separate rooms.
- Have group A rank the problem list by benefit to the organization.
- Have group B rank the problems by how much effort will be required to solve them.
- Set up the problem priority matrix as shown in Figure 6.3. The extended number is arrived at by multiplying the benefit ranking by the effort ranking for each problem.

The lowest extended number shown in Figure 6.3 is the problem that should be solved first. Solving it will yield the most benefit to the organization with the least expenditure of effort. Other problems will yield more benefit or will require less effort. But when both benefit and effort required are taken into account, the lowest extended number would be the first to be solved.

6. *Gather information about the problem.* When the problems have been prioritized, the temptation will be to jump right in and begin solving them. This can be a mistake. A better approach is to collect all available information about a problem before pursuing solutions. Two kinds of information can be collected: objective and subjective. Objective information is factual. Subjective information is open to interpretation.

Problem	Ranking by Benefit	Ranking by Effort	Extended No. (Multiplied)	Final Ranking
A	2	5	10	2
B	6	3	18	4
C	1	9	9	1
D	8	2	16	3
E	10	1	10	2
F	3	8	24	6
G	5	4	20	5
H	4	10	40	7
I	7	7	49	8
J	9	6	54	9

FIGURE 6.3   Problem priority matrix. Lowest number = highest priority

Rarely will the information collected be only objective in nature. Nothing is wrong with collecting subjective information, as long as the following rules of thumb are adhered to for both objective and subjective information:

- Collect only information that pertains to the problem in question.
- Be thorough (it's better to have too much information than too little).
- Don't waste time re-collecting information that is already on file.
- Allow sufficient time for thorough information collection, but set a definite time limit.
- Use systematic tools such as Pareto charts, histograms, and cause-and-effect diagrams.

Specifying the Problem

Specifying the problem means breaking it down into its component parts. All problems can be broken into five basic components as follows:

1. Who is the problem affecting?
2. What is the problem? This is a restatement of the final results of the why-why process.
3. When did the problem occur first, and when did it occur last? How often is it occurring?

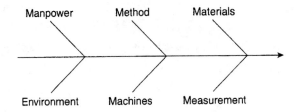

FIGURE 6.4   Sample cause-and-effect diagram.

4. Where does the problem occur? The answer to this question should be specific. If a machine in Department A is the problem, where should be specified not just as Department A, but also where in the machine the problem occurs (in the electrical system, in the hydraulic system, etc.).

5. How much? What is the extent of the problem? Are there defective parts in every 50 produced? Are there two breakdowns per shift? This question should be answered in quantifiable terms whenever possible.

Identifying Causes

Identifying causes is a critical step in the process. It involves the pairing off of causes and effects. Effects are the problems that have already been identified. After one such problem has been targeted for solving, an effective tool for isolating the causes of the problem is the cause-and-effect, or fishbone, diagram. A cause-and-effect diagram is illustrated in Figure 6.4. The six "spines" on this particular diagram represent the six major groupings of causes: manpower (personnel), method, materials, machines (equipment), measurement, and environment. All causes of workplace problems fall into one of these major groupings. Using the diagram, team members brainstorm causes under each grouping. For example, under the machine grouping, a cause might be insufficient maintenance. Under the manpower grouping, a cause might be insufficient training. The key is to isolate the root cause.

Isolating the Root Cause

To isolate the most probable root cause or causes, each cause identified on the cause-and-effect diagram is compared with the problem specifications developed earlier (i.e., who, what, when, where, how much).

When comparing potential causes and problem specifications for each cause, team members will have three possibilities: the cause will fully explain the specification, the cause will partially explain the specification, or the cause will not explain the specification. A cause that can fully explain the specification is a likely candidate to be the root cause. If more than one

Problem Definition	Problem Solution (Desired Result)
▪ Job 21A is over budget by 22%.	▪ Job 21A must be brought under budget by 2%.
▪ Reject rate is too high by 12%.	▪ Reject rate must be brought to 0%.

FIGURE 6.5 Problem definitions and problem solutions.

cause fully explains the specification, there may be more than one cause of the problem.

Finding the Optimum Solution

With the problem and its most probable root cause identified, the next step is to find the optimum solution, or the desired result. The first task in this step is to develop a solution definition that clearly explains the effect the solution is to have. Figure 6.5 gives examples of problem definitions and their corresponding solution definitions.

With the definition in place, the team brainstorms possible solutions and creates a list. Perry Johnson, Inc., has developed a tool known as SCAMPER that can improve effectiveness in building a solutions list:[2]

- *Substituting.* Can the problem be solved by substitution? Can a new process be substituted for the old? Can one employee be substituted for another? Can a new material be substituted for the old?
- *Combining.* Can the problem be solved by combining two or more tasks, processes, activities, operations, or other elements?
- *Adapting.* Can the problem be solved by adapting an employee, a process, a product, or some other element to another purpose?
- *Modifying.* Can the problem be solved by modifying a process, job description, design, or something else?
- *Putting to other uses.* Can the problem be solved by putting a resource to other uses?
- *Eliminating.* Can the problem be solved by eliminating a position, part, process, machine, product, or something else?
- *Replacing.* Can the problem be solved by replacing an individual, part, process, machine, product, or something else?

With the solutions list developed, the next step is to identify the optimum entry on the list. One way to do this is by group consensus. A more

Categories of Costs/Benefits	Purchase New Machines	Retrofit/ Automate	Add a Second Shift
Costs			
Personnel	$ -0-	$ -0-	$180,000
Equipment	95,000	26,000	-0-
Downtime	-0-	11,400	-0-
Installation	15,200	7,600	-0-
Training	-0-	15,200	16,300
	$110,200	$ 60,200	$196,300
Benefits			
Reduced overhead	$ 27,500	$ 27,500	$ 27,500
Elimination of late fees	32,000	32,000	32,000
Elimination of canceled orders . .	75,000	75,000	75,000
	$134,500	$134,500	$134,500
Cost-to-benefit ratio	1 to 1.2	1 to 2.23	1.46 to 1

FIGURE 6.6   Cost-benefit matrix.

objective approach is to undertake a cost-to-benefit analysis. This involves setting up a cost-benefit matrix. The matrix should contain cost categories and show the actual dollar costs in each category for each potential solution. The matrix is repeated to show the dollar contribution for each category of benefits. Then the total cost of each solution is compared with the total benefit to derive a cost-to-benefit ratio for each potential solution.

Figure 6.6 shows a cost-benefit matrix based on three proposed solutions to the following problem: the plastic coating unit is backlogged by two months. The three potential solutions are as follows:

1. Purchase two additional plastic coating machines.
2. Retrofit existing plastic coating machine with computer controls and automate the process.
3. Add a second shift and run the plastic coating unit an additional 8-hour shift each day.

The cost-benefit matrix in Figure 6.6 shows the total cost for each potential solution. The costs range from a low of $60,200 to a high of $196,300. The dollar value in benefits reveals that adding a second shift will produce costs that outweigh the benefits. The other two options are both feasible. However, the retrofit and automate option has a better cost-to-benefit ratio. With this option, every dollar spent will produce $2.23 in benefits. The cost-to-benefit analysis indicates that retrofitting and automating the existing process is the optimum solution.

Implementing the Optimum Solution

The implementation phase of the process is critical. If handled properly, the problem will be solved in a way that improves the process. However, if implementation is not handled properly, new and even more serious problems can be created.

The key to effective implementation of a solution is to take a systematic approach. Perry Johnson recommends developing an action plan with the following components:

- Actions to be taken
- Methods for taking each action
- Resources needed for each action
- Special needs for each action
- Person responsible for each action
- Deadline for each action

DECISION MAKING FOR EFFECTIVE LEADERSHIP

All people make decisions. Some are minor: What should I wear to work today? What should I have for breakfast? Some are major: Should I accept a job offer in another city? Should I buy a new house? Regardless of the nature of the decision, decision making can be defined as follows:

> Decision making is the process of selecting one course of action from among two or more alternatives.

Decision making is a critical task for leaders. Decisions play the same role in an organization that fuel and oil play in an automobile engine: they keep it running. The work of an organization cannot proceed until decisions are made.

Leadership Tip

"Creativity is so delicate a flower that praise tends to make it bloom, while discouragement often nips it in the bud. Any of us will put out more and better ideas if our efforts are appreciated."[3]

—Alexander F. Osborn
American Business Executive

Consider the following example. Because a machine is down, the production department at DataTech, Inc., has fallen behind schedule. With this machine down, DataTech cannot complete an important contract on time without scheduling at least 75 hours of overtime. The production manager faces a dilemma. On the one hand, no overtime was budgeted for the project. On the other hand, there is substantial pressure to complete this contract on time because future contracts with this client may depend on it. The manager must make a decision.

In this case, as in all such situations, it is important to make the right decision. But how do leaders know when they have made the right decision? In most cases there is no single right choice. If there were, decision making would be easy. Typically several alternatives exist, each with its own advantages and disadvantages. In the DataTech example the manager has two alternatives: authorize 75 hours of unbudgeted overtime or risk losing future contracts. If the manager authorizes the overtime, his company's profit for the project will suffer, but its relationship with the client may be protected. If the manager refuses to authorize the overtime, the company's profit on this project will be protected, but the relationship with this client may be damaged. These and other types of decisions must be made all the time in the modern workplace.

Leaders should be prepared to have their decisions evaluated, questioned, and even criticized after the fact. Although it may seem unfair to conduct a retrospective critique of decisions that were made during the "heat of battle," having one's decisions evaluated is part of accountability, and it can be an effective way to improve a leader's decision-making skills.

There are two ways to evaluate decisions. The first is to examine the results. In every case where a decision must be made, there is a corresponding result. That result should advance an organization toward the accomplishment of its goals. To the extent that it does, the decision is usually considered a good one.

Leaders have traditionally had their decisions evaluated by results. However, this is not the only way that decisions should be evaluated. Regardless of results, it is wise also to evaluate the process used in making a decision. Positive results can cause a manager to overlook the fact that a faulty process was used, and in the long run a faulty process will lead to negative results more frequently than to positive ones.

For example, suppose a manager must choose from among five alternatives. Rather than collect as much information as possible about each, the manager chooses randomly. She has one chance in five of choosing the best alternative. Such odds occasionally produce a positive result, but typically they don't. This is why it is important to examine the process as well as the result, not just when the result is negative but also when it is positive.

Leadership Profile   **Mike Armstrong and Decision Making at AT&T[4]**

AT&T is one of the largest and most recognized communications companies in the world. The company has more than 80 million customers—a customer base larger than many countries in the world. AT&T's services include wireless telephony, local telephone service, Internet access, and international telephone service.[5]

AT&T transformed itself from being just a long-distance, point-to-point communications company into a diversified, comprehensive, global provider of "any distance" communications. The person who led this impressive transformation was Mike Armstrong. One of the most important leadership abilities Armstrong brought to the CEO's position was the ability to make a decision. His first decision was to establish a new vision for AT&T. Armstrong's vision was to make AT&T such a comprehensive provider of communications services that customers would need no other provider.[6] On the surface Armstrong's decision might sound like a no-brainer, but consider what making this decision really meant.

Consider the magnitude of the risk Armstrong took in making the investments that would be necessary to transform his company: (a) $12 billion to purchase Teleport Communication; (b) $5 billion to purchase a global network from IBM; (c) $48 billion to purchase TCI (the second largest cable company in the United States); and (d) $54 billion to purchase MediaOne Group.[7] Obviously, Armstrong decided to spend a lot of money, but his philosophy of leadership is that it is the leader's job to make big decisions. Leaders who cannot make decisions of this magnitude cannot lead companies of AT&T's magnitude.

To paraphrase Armstrong's thoughts on decision making is to say that in the final analysis, it's about courage. You must have the courage to make the tough decisions. There is never enough time to consider every minute detail of every possible option and there is never sufficient data about the costs versus the benefits of those options. Consequently, a leader has to get the most information and the best information available, apply his instincts, and make the decision.[8]

THE DECISION-MAKING PROCESS

Decision making is a process. For the purpose of this book, the decision-making process is defined as follows:

> The decision-making process is a logically sequenced series of activities through which decisions are made.

Numerous decision-making models exist. Although they appear to have major differences, all involve the following steps, shown in Figure 6.7.

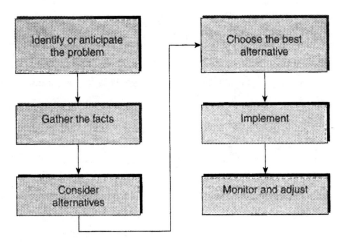

FIGURE 6.7   Decision-making model.

Identify or Anticipate the Problem

If leaders can anticipate problems, they may be able to prevent them. Anticipating problems is like driving defensively: never assume anything. Look, listen, ask, and sense. For example, if you hear through the grapevine that a team member's child has been severely injured and hospitalized, you can anticipate the problems that may occur. She is likely to be absent, or if she does come to work, her pace may be slowed. The better that leaders know their employees, customers, technological systems, products, and processes, the better able they will be to anticipate problems.

Gather the Facts

Even the most perceptive leaders will be unable to anticipate all problems or to understand intuitively what is behind them. For example, suppose a manager notices a "Who cares?" attitude among team members. This manager might identify the problem as poor morale and begin trying to improve it. However, he would do well to gather the facts first to be certain of what is behind the negative attitudes. The underlying cause or causes could come from a wide range of possibilities: an unpopular management policy, dissatisfaction with the team leader, a process that is ineffective, problems at home, and so forth. Using the methods and tools described earlier in this chapter, the manager should separate causes from symptoms and determine the root cause of the problem. Only by doing so will the problem be permanently resolved.

The factors that might be at the heart of a problem include not only those for which a leader is responsible (policies, processes, tools, training, personnel assignment, etc.) but also ones beyond the leader's control (personal matters, regulatory requirements, market and economic influences, etc.). For those falling within the leader's range of authority, the leader must make sound, informed decisions based on facts. For the others, the organization has to adapt.

Consider Alternatives

Considering the alternatives involves two steps: (1) list all of the various alternatives available, and (2) evaluate each alternative in light of the facts. The number of alternatives identified in the first step will be limited by several factors. Practical considerations, the leader's range of authority, and the cause of the problem will all limit a manager's list of alternatives. After the list has been developed, each entry on it is evaluated. The main criterion against which alternatives are evaluated is the desired outcome. Will the alternative being considered solve the problem?

Cost is another criterion used in evaluating alternatives. Different alternatives have different costs, which might be measured in financial terms, by employee morale, by the organization's image, or by a client's goodwill. Such costs should be considered when evaluating alternatives. In addition to applying objective criteria and factual data, leaders also need to apply their judgment and experience when considering alternatives.

Choose the Best Alternative; Implement, Monitor, and Adjust

After all alternatives have been considered, one must be selected and implemented, and after an alternative has been implemented, leaders must monitor progress and adjust appropriately. Is the alternative having the desired effect? If not, what adjustments should be made? Selecting the best alternative is never a completely objective process. It requires study, logic, reason, experience, and even intuition. Occasionally, the alternative chosen for implementation will not produce the desired results. When this happens and adjustments are not sufficient, it is important for leaders to cut their losses and move on to another alternative.

Leaders should avoid falling into the ownership trap. This happens when they invest so much ownership in a given alternative that they refuse to change even when it becomes clear the idea is not working. This can happen at any time but is more likely to happen when a leader selects an alternative that runs counter to the advice the leader has received, is unconventional, or is unpopular. The leader's job is to solve the problem. Showing too much ownership in a given alternative can impede the ability to do so.

EMPLOYEE INVOLVEMENT IN DECISION MAKING

Employee involvement and empowerment can improve decision making. Employees are more likely to show ownership in a decision they helped make. Correspondingly, they are more likely to support a decision for which they feel some ownership. There are many advantages to be gained from involving employees in decision making. There are also factors that, if not understood and properly handled, can lead to problems.

Advantages of Employee Involvement

Involving employees in decision making can result in a more accurate picture of what the problem really is and a more comprehensive list of potential solutions. It can help managers do a better job of evaluating alternatives and selecting the best one to implement.

Perhaps the most important advantages are gained after the decision is made. Employees who participate in the decision-making process are more likely to understand and accept the decision and have a personal stake in making sure that the alternative selected succeeds.

Potential Problems with Employee Involvement

Involving employees in decision making can also lead to problems. The major potential problem is that it takes time, and managers do not always have time. Other potential difficulties are that it (a) takes employees away from their jobs and (b) can result in conflict among team members. Next to time, the most significant potential problem is that employee involvement can lead to democratic compromises that do not necessarily result in the best decision. In addition, disharmony may result when a decision maker rejects the advice of the group. Nevertheless, if care is taken, leaders can gain all of the advantages while avoiding the potential disadvantages associated with employee involvement in decision making.

CREATIVITY IN DECISION MAKING

The increasing pressures of a competitive marketplace are making it more important for organizations to be flexible, innovative, and creative in decision making. To survive in an unsure, rapidly changing marketplace, organizations must be able to adjust rapidly and change direction quickly. To do so requires creativity at all levels of the organization.

Creativity Defined

Like leadership, creativity has many definitions, and viewpoints vary about whether creative people are born that way or develop the ability over time. For the purposes of this book, creativity can be viewed as an approach to problem solving and decision making that is imaginative, original, and innovative. Developing such perspectives requires that decision makers have knowledge and experience regarding the issue in question.

Creative Process

According to H. Von Oech, the creative process proceeds in four stages: preparation, incubation, insight, and verification.[9]

- *Preparation* involves learning, gaining experience, and collecting and storing information in a given area. Creative decision making requires that the people involved be prepared.
- *Incubation* involves giving ideas time to develop, change, grow, and solidify. Ideas incubate while decision makers get away from the issue in question and give the mind time to sort things out. Incubation is often a function of the subconscious mind.
- *Insight* follows incubation. It is the point at which a potential solution falls into place and becomes clear to the decision makers. This point is sometimes seen as a moment of inspiration. However, inspiration rarely occurs without having been preceded by perspiration, preparation, and incubation.
- *Verification* involves reviewing the decision to determine whether it will actually work. At this point, traditional processes such as feasibility studies and cost-benefit analyses are used.

Factors That Inhibit Creativity

Numerous factors can inhibit creativity. Following are some of the more prominent.[10]

- *Looking for just one right answer.* Seldom is there just one right solution to a problem.
- *Focusing too intently on being logical.* Creative solutions sometimes defy logic and conventional wisdom.
- *Avoiding ambiguity.* Ambiguity is a normal part of the creative process. This is why the incubation step is so important.
- *Avoiding risk.* When organizations don't seem to be able to find a solution to a problem, it often means decision makers are not willing to give an idea a chance.
- *Forgetting how to play.* Adults sometimes become so serious they forget how to play. Playful activity can stimulate creative ideas.
- *Fear of rejection or looking foolish.* Nobody likes to look foolish or feel rejection. This fear can cause people to hold back what might be creative solutions.
- *Saying "I'm not creative."* People who decide they are not creative won't be. Any person can think creatively and can learn to be even more creative.

Helping People Think Creatively

In this age of technology and global competition, creativity in decision making and problem solving is critical. Although it is true that some people are naturally more creative than others, it is also true that any person can learn to think creatively. In the modern workplace, the more people who think creatively, the better. Darrell W. Ray and Barbara L. Wiley recommend the following strategies for helping employees think creatively:[11]

■ *Idea vending.* This is a facilitation strategy. It involves reviewing literature in the field in question and compiling files of ideas contained in the literature. These ideas are periodically circulated among employees as a way to get them thinking. This will facilitate the development of new ideas by the employees. Such an approach is sometimes called stirring the pot.

■ *Listening.* One of the factors that causes good ideas to fall by the wayside is poor listening. Leaders who are perpetually too hurried to listen to employees' ideas do not promote creative thinking. On the contrary, such managers stifle creativity. In addition to listening to employees' ideas, good and bad, managers should listen to the problems employees discuss in the workplace. Each problem is grist for the creativity mill.

■ *Idea attribution.* A leader can promote creative thinking by subtly feeding fragments of ideas to employees and encouraging them to develop the idea fully. When an employee develops a creative idea, she gets full attribution and recognition for the idea. Time may be required before this strategy pays off, but with patience and persistence it can help employees become creative thinkers.

LESSONS ON PROBLEM SOLVING AND DECISION MAKING FROM SELECTED LEADERS

Following are excerpts from the lives of several leaders that exemplify some of the problem-solving and decision-making principles set forth in this chapter. The leaders selected for inclusion are Colin Powell, Robert Noyce, Andrew Grove, Gordon Moore, and Margaret Thatcher.

Colin Powell[12]

"A career military man, Colin Powell came to broad public attention as chairman of the Joint Chiefs of Staff during the Persian Gulf War of 1990–91 and became sufficiently popular to contemplate a run for the White House. In 2001 he became the nation's first African American secretary of state."[13] Although his greatest achievements may be yet to come, Colin Powell has already achieved more than most men even dream of. His life and career thus far teach many valuable leadership lessons about problem solving and decision making. Prominent among these lessons are the following:

Effective leaders are willing to innovate and improvise.

Effective leaders accept personal responsibility for their decisions but recognize the necessity of coordinating the efforts of many different stakeholders.

Colin Powell rose from humble beginnings to achieve top military command and high governmental office. He was the first African American to serve as chairman of the Joint Chiefs of Staff. He had overall responsibility for Operation Just Cause, the removal of Panamanian dictator Manuel Noriega from office, an undertaking that required both innovative problem solving and enormous diplomatic skills. Perhaps his greatest moment as an innovative problem solver and decision maker came when he oversaw the Persian Gulf War, one of the most diplomatically and logistically difficult military operations ever undertaken. Following his success in leading the U.S. military in the war, Powell became the first African American secretary of state.

Powell's challenges during the Persian Gulf War would have seemed insurmountable to a lesser leader. To remove the military forces of Iraqi dictator Saddam Hussein from the tiny country of Kuwait, Powell had to coordinate a coalition of 48 nations, manage the deployment of more than half a million troops, and supervise the logistics of bringing to bear in the right place at the right time a huge armada of ships and airplanes carrying a massive amount of equipment and supplies.

"Working closely with field commanders such as General H. Norman Schwarzkopf, Powell directed a lightning war, the ground phase of which was completed in about 100 hours. At a cost of 95 troops killed, 368 wounded, and 20 missing in action, the coalition inflicted about 160,000 Iraqi casualties, including 50,000 killed, 50,000 wounded, and some 60,000 taken prisoner. Vast quantities of Iraqi arms and equipment were destroyed."[14] The example of Colin Powell is one that could well be emulated by technical professionals who aspire to leadership roles.

Robert Noyce, Andrew Grove, and Gordon Moore[15]

Three of the most important names in the computer revolution are Robert Noyce, Andrew Grove, and Gordon Moore. These are the leaders principally responsible for establishing Intel Corporation and for making it the world's largest microprocessor manufacturer. "For the head of a $16.2 billion company responsible for 26,000 employees, Andrew Grove, the chief executive officer of Intel Corporation, still has the look of an entrepreneur."[16] Grove thinks decision making is critical and leaders in business must be willing to make decisions. A good decision is, of course, best. But even making a wrong decision is better than hedging.

"Intel has never hedged. From the beginning it has forged relentlessly into new territory. In 1968, when Gordon Moore and Robert Noyce left the security of a large, established firm to start their own company, their plan was to manufacture a product they had yet to invent: a tiny semiconductor chip with the same capacity to store computer memory as the large magnetic cores used in mainframe computers. Under the direction of Moore and Noyce, Intel's engineers set out to pack more and more computing power on ever-smaller chips. In 1971 they made a chip that could revolutionize the operation of the computer. The microprocessor, as it was called, is a device now ranked with McCormick's reaper and Henry Ford's assembly line as a milestone in the history of invention." The decision-making and problem-solving lesson taught by the careers of Robert Noyce, Andrew Grove, and Gordon Moore is as follows:

> To lead their organizations into new frontiers, leaders must be willing to innovate and think creatively when making decisions.

"The invention of the microprocessor was simply the beginning. Intel, the early technological leader, has made a strenuous effort to maintain its lead. . . . Even after establishing its microprocessors, which are produced in state-of-the-art factories around the world, as the industry standard, Intel continues to operate as if it were a research institution. In recent years its annual budget for research and development has topped $1 billion."[17]

Research and development coupled with innovation have been Intel's hallmarks over the years. Noyce, Grove, and Moore did not just invent a new product; they invented a new industry. Under their leadership, Intel not only developed an ongoing succession of new computer chips but also developed the means to produce each new generation of chips. Developing a means to produce each successive generation of chips was one of Intel's most significant innovations.

"On an old-fashioned transistor, an impurity the size of, say, a cookie crumb, would interfere with performance. Next to a transistor measuring less than one micron, a germ—a single bacterium—looks like a boulder and renders the whole chip worthless. Intel had to design production rooms in which all the air was filtered every few seconds, leaving less than one such particle per cubic foot. Humans, those roving dust storms of dandruff, viruses, spit, and lint, had to be sealed inside special suits in order to work in *clean rooms*."[18]

Intel's innovations went beyond just developing the microchip, something that in and of itself ranks near the top of the list of the most significant technological innovations of all time. The company also pioneered the means of producing microchips. In addition to the innovative clean room concept, this also involved finding locations around the world in which the earth is sufficiently flat, stable, and free of vibration. The slightest tremor, even one that could not be felt by human beings, could undermine the production process. Intel now has chip-manufacturing plants in Israel and Ireland, two countries with especially stable ground.

Margaret Thatcher[19]

Margaret Thatcher could have been sure of her place in history even if she had been a mediocre leader. As Great Britain's first woman prime minister, her place in history was secure the day she was elected. But Margaret Thatcher, the daughter of a grocer, was anything but mediocre, and her gender was among her least distinguishing features. Margaret Thatcher's leadership of the United Kingdom transcended gender. Her time in office—an era that came to be known as the "Thatcher Revolution"—lasted longer than that of any other British prime minister in the 20th century. Not only did Margaret Thatcher change the course of social welfare and reverse a downwardly spiraling economy in the United Kingdom, but she led the country successfully through the Falkland Islands War in 1982.[20] The decision-making and problem solving lessons taught by the career of Margaret Thatcher are as follows:

> Leaders must be willing to make difficult decisions that may not be popular, at least not in the short run.

> Leaders must be willing to be responsible and accountable for the results of their decisions.

When Thatcher took over as prime minister, Britain's economy was in terrible shape and getting worse daily. Thatcher knew what had to be done, but she also knew that doing it would make her the target of vicious political attacks by opponents. Nevertheless, Thatcher forged ahead with her program of social and economic reforms. She reduced welfare programs, privatized government agencies and functions, and took a strong stand against the labor unions that she believed were dragging down the country's industrial productivity and quality, as well as its ability to compete in the global marketplace. She even instituted a poll tax that applied to all would-be voters regardless of socio-economic status. Thatcher was condemned by some as a heartless conservative and by others as undemocratic and oppressive. However, she was successful in reversing Britain's downward economic spiral. By the time Thatcher stepped down as prime minister, Britain was re-emerging as a major player in the global marketplace—something that had not been the case for more than two generations.[21]

Margaret Thatcher's career illustrates conclusively how important it is for leaders to look at the big picture, have clear goals, and make decisions that, although controversial, are in the long-term best interests of the organization. Her tenure as Great Britain's political leader proves a management adage concerning difficult decisions: Nothing succeeds like success.

Leadership Tip

"I sometimes believe that large corporations are not capable of real innovation. The giants . . . did not become a factor in the semiconductor industry even though they did spend most of the money on early development. They developed a good technical base, but small companies picked that up and made the applications."[22]

—W.A. Pieczonka
President, Linear Technologies, Inc.

Summary

1. *Decision making* is a process of selecting one course of action from among two or more alternatives. Decisions should be evaluated not just by results but also by the process used to make them.

2. A *problem* is a situation in which what exists does not match what is desired; or, put another way, the discrepancy between the current and the desired state of affairs. Problem solving is not about putting out fires. It is about continual improvement. An effective problem-solving model is the Perry Johnson method.

3. The decision-making process is a logically sequenced series of activities through which decisions are made. These activities include identifying or anticipating the problem; gathering relevant facts; considering alternative solutions; choosing the best alternative; and implementing, monitoring, and adjusting the alternative chosen. All approaches to decision making are objective, subjective, or a combination of the two.

4. *Creativity* is an approach to problem solving and decision making that is imaginative, original, and innovative. The creative process proceeds in four stages: preparation, incubation, insight, and verification. Factors that inhibit creativity include looking for just one right answer, being too logical, avoiding ambiguity, avoiding risk, forgetting how to play, fearing rejection, and saying "I'm not creative." Three strategies for helping people think creatively are idea vending, listening, and idea attribution.

Key Terms and Concepts

Cause-and-effect (fishbone)
  diagram
Creativity
Decision making

Decision-making process
Idea attribution
Idea vending
Incubation

Insight

Objective decision making

Perry Johnson method

Preparation

Problem-solving team

SCAMPER

Subjective decision making

Why-why

Review Questions

1. Discuss decision making as it relates to total quality.
2. Explain how to evaluate decisions.
3. Describe the Perry Johnson method for problem solving.
4. Define the decision-making process and explain each step in it.
5. Contrast and compare objective and subjective decision making.
6. What are the advantages and disadvantages of employee involvement in decision making?
7. Explain creativity as a concept and the role it can play in decision making.

LEADERSHIP SIMULATION CASES

The following simulations are provided to generate additional thought and discussion about the principles of leadership explained in this chapter. Readers are encouraged to consider how the situations presented in these cases might apply to them and to discuss the cases with other leaders and leadership candidates.

CASE 6.1 I Need a Decision-Making Process

Jim Bolton had been an electrical engineer with Southern Technologies, Inc. (STI) for 20 years when he was promoted to the position of vice president for engineering. Now that he is leading a major division, Bolton is finding that decision making is much more difficult than when he was a design engineer. Speaking to a colleague, a frustrated Bolton said, "When I was working in design, decisions were easy. I knew my product, I knew our processes, and I knew what our customers wanted. Consequently, when I had to make a decision, it was easy. I didn't need to research the question, ask for the opinions of other people, or consider a hundred different seemingly unrelated issues. I was the expert, and I knew what to do."

Bolton went on to explain that things had changed radically since he stepped up to the executive leadership level. "Now I am no longer the expert. I have to make decisions all the time that are of critical importance, and I might know little or nothing about the issue in question. What I need is a process for making decisions on some systematic basis, rather than just on intuition or foreknowledge."

Discussion Questions

1. Have you ever had to make a decision on a matter you knew little or nothing about? Explain.
2. Explain the decision-making process you would recommend to Bolton if you were the colleague he was talking to.

CASE 6.2 You Have to Identify the Root Cause of the Problem

"I have solved this problem five times in the past two months and it keeps coming back!" Jane Corbel, manufacturing director for FWB Manufacturing, Inc., was alone in her office, so nobody heard her exclamation of frustration. The all-important XL-12 project was both behind schedule and over budget. And Corbel knew why, or at least she thought she did—machine breakdowns. It seemed the machines in her shop could not run even one full shift without experiencing some type of breakdown. The problem, as Corbel saw it, was complicated by the fact that every time a machine broke down there was a different reason. "This situation is going to drive me crazy," snapped the frustrated manager to herself.

Out of frustration, Corbel decided to pay a visit to an old friend of hers. Dr. Tom Newland had been more than Corbel's major professor when she had majored in operations management at West Florida University; he had been her mentor and benefactor. It was Newland who convinced the CEO of FWB Manufacturing to give the inexperienced Corbel a chance to prove what she could do as a manager of the company's machining unit.

"Tom, I am at a loss. My machines keep breaking down—different problems every time. The vendor has sent service people out promptly every time we have a breakdown. Consequently, we always get up and running pretty quickly after a breakdown. But just as soon as one machine is running, another breaks down," said Corbel. "What do you think the problem is?" asked the professor. "At first I thought it was just faulty machines, but the vendor claims that no other companies using the same machines are having these problems. In fact, I urged my boss to invest in these machines because of their reputation for quality."

Corbel went on to explain that she had even tried to pinpoint the problem herself using the trouble shooting checklist provided by the vendor. "It sounds like you are just guessing to me, Jane," said Newland. "You need to identify the root cause of your problem." "I know," responded Corbel. "That's why I'm here. You taught us how to go about identifying the root cause of a problem, but I need a quick review."

Discussion Questions

1. Have you ever had a problem that you thought you solved, but it kept coming up in spite of your efforts? Explain.
2. Put yourself in Dr. Newland's place. What would you tell Corbel in the way of giving her a quick review of how to identify the root cause of problems such as the one she is experiencing with her machines?

CASE 6.3 What We Need Is Some Creativity

John Marshall III had called a special meeting of his executive leadership team to discuss his company's future. "It's simple, folks. Either we find some new and different ways to do business, or we are going to go out of business," claimed Marshall to open the meeting. He could see that his statement had the desired effect.

Marshall had inherited his company, Marshall Technical Enterprises, from his father, who had inherited it a generation earlier from his father. The grandfather had established the company as an engineering and manufacturing company that produced farm implements. The father had grown the company by branching out into a broader range of farming-related equipment. Now the grandson, John Marshall III, was afraid that he might be the unfortunate CEO to oversee the dissolution of the venerable family business.

"In a nutshell, gentlemen, foreign competition is killing us. Our major customers are dropping us like a hot potato and migrating to competitors overseas—mostly Japanese firms," said Marshall. "If we can't come up with some better ideas than we have right now, I am going to have to put this company on the market. And, frankly, if I am forced to do that, I'll be lucky to get half of what I think it's worth."

"What we need is some creative thinking," said the company's chief operating officer. "And I don't mean among just the few of us sitting in this room right now. I mean we need some good creative thinking from all employees at all levels." "Indeed we do," admitted Marshall. "But how are we going to get it? That's the question."

Discussion Questions

1. Have you ever worked in a situation in which marketplace circumstances had changed, but the company kept doing things the same old way? Explain.
2. If you were present in the meeting John Marshall called in this case, what recommendations would you make for engendering creative thinking at all levels of the company?

Endnotes

[1] Perry L. Johnson, Rob Kantner, and Marcia A. Kikora, *TQM Team-Building and Problem-Solving* (Southfield, MI: Perry Johnson, 1990), Chapter 1, pp. 10–15.

[2] Perry L. Johnson, Rob Kantner, and Marcia A. Kikora, 11–12.

[3] Louis E. Boone, *Quotable Business*, 2nd ed. (New York: Random House, 1999), 12.

[4] Thomas J. Neff and James M. Citrin, *Lessons from the Top* (New York: Currency/Doubleday, 2001), 35–40.

[5] Thomas J. Neff and James M. Citrin, 39.

[6] Thomas J. Neff and James M. Citrin, 37.

[7] Thomas J. Neff and James M. Citrin, 37.

[8] Thomas J. Neff and James M. Citrin, 37.

[9] H. Von Oech, *A Whack on the Side of the Head* (New York: Warner, 1983), 77.

[10] Von Oech, 77.

[11] Darell W. Ray and Barbara L. Wiley, "How to Generate New Ideas," in *Management for the 90s: A Special Report from* Supervisory Management (New York: AMACOM, 1991), 6–7.

[12] Alan Axelrod, *Profiles in Leadership* (Upper Saddle River, NJ: Prentice Hall, 2003), 429–432.

[13] Alan Axelrod, 429.

[14] Alan Axelrod, 431.

[15] Daniel Gross, *Forbes Greatest Business Stories of All Time* (New York: John Wiley & Sons, Inc., 1996), 247–265.

[16] Daniel Gross, 247.

[17] Daniel Gross, 247–248.

[18] Daniel Gross, 263.

[19] Alan Axelrod, 514–517.

[20] Alan Axelrod, 514.

[21] Alan Axelrod, 515–516.

[22] Louis E. Boone, 92.

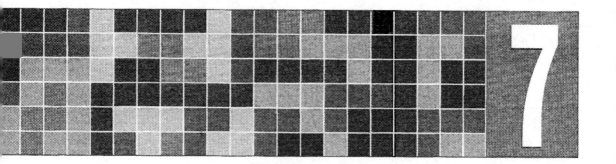

Be a Positive Change Agent

"In order to survive in a competitive environment, organizations must improve continually. This means they have to change continually. Organizations that fail to change—whether from complacency, ignorance, or conceit—are doomed to mediocrity, or worse yet, failure."

—The Author

OBJECTIVES

- Explain the importance of leadership in achieving positive change.
- Summarize the representative views of managers and employees regarding change.
- Explain the relationship of restructuring and change.
- Describe the process for leading change.
- Summarize the change-related lessons of selected leaders.

Few things call more clearly for effective leadership than organizational change. It takes an effective leader to overcome an inherent characteristic of organizations—organizational inertia. Students of physics know that inertia refers to the tendency of a body at rest to stay at rest until sufficient force is applied to break the inertia and get it moving. It also refers to the tendency of a body in motion to continue moving until sufficient force is applied to stop it.

Organizations are like the "body" referred to in this principle. An organization with a static culture (a body at rest) will tend to remain static until sufficient leadership is applied to get it moving. The good news is that, like the body referred to in physics, a dynamic organization will tend to stay dynamic unless sufficient force is applied to stop it. Another way to state this principle as it applies to organizations is

> An organization's culture, once established, will tend to perpetuate itself until sufficient leadership is applied to change it.

How to effectively apply leadership to organizations to ensure that they not only change to stay current but also get out in front of change is the subject of this chapter.

IMPORTANCE OF LEADERSHIP IN CHANGE

In a competitive and rapidly changing marketplace, technology companies are constantly involved in the development of strategies for keeping up, staying ahead, or setting new directions. What can managers do to play a positive role in the process? David Shanks recommends the following strategies:[1]

- Have a clear vision and corresponding goals.
- Exhibit a strong sense of responsibility.
- Be an effective communicator.
- Have a high energy level.
- Have the will to make change happen.

Shanks developed these strategies to help executives guide their companies through periods of corporate stress and change, but they also apply to other personnel. The characteristics of good leaders apply to personnel at any level who must help their organization deal with the uncertainty caused by change.

FACILITATING CHANGE AS A LEADERSHIP FUNCTION

The following statement by management consultant Donna Deeprose carries a particularly relevant message for technical professionals who would lead:

In an age of rapidly accelerating technology, restructuring, repositioning, downsizing, and corporate takeovers, change may be the only constant. Is there anything you can do about it? Of course there is. You can make change happen, let it happen to you, or you can stand by while it goes on around you.[2]

Deeprose divides people into three categories based on how they handle change: driver, rider, or spoiler.[3] People who are drivers lead their organizations in new directions as a response to change. People who just go along, reacting to change as it happens rather than getting out in front of it, are riders. Managers who actively resist change are spoilers. Deeprose gives examples of how a driver would behave in a variety of situations:[4]

- In viewing the change taking place in an organization, drivers stay mentally prepared to take advantage of the change.
- When facing change about which they have misgivings, drivers step back and examine their own motivations.
- When a higher manager has an idea that has been tried before and failed, drivers let the boss know what difficulties were experienced earlier and offer suggestions for avoiding the problems this time.
- When a company announces major changes in direction, drivers find out all they can about the new plans, communicate what they learn with employees, and solicit input to determine how to make a contribution to the achievement of the company's new goals.
- When a change will affect other departments, drivers go to these other departments and explain the change in their terms, solicit their input, and involve them in the implementation process.
- When demand for their unit's work declines, drivers solicit input from users and employees as to what modifications and new products or services might be needed and include the input in a plan for updating and changing direction.
- When an employee suggests a good idea for change, drivers support the change by justifying it to higher management and using their influence to obtain resources for it while countering opposition to it.
- When their unit is assigned a new, unfamiliar task, drivers delegate the new responsibilities to their employees and make sure they get the support and training needed to succeed.

These examples show that a driver is a manager who exhibits the leadership characteristics necessary to play a positive, facilitating role in helping employees and organizations successfully adapt to change on a continual basis.

PERCEPTIONS OF EMPLOYEES AND MANAGERS ON CHANGE

One of the difficulties organizations face when attempting to facilitate change is the different perceptions of employees and managers concerning change. Employees often view change as something done to them. Managers often regard it as something done in spite of employees who just won't cooperate. The following quote is representative of how employees sometimes feel about change.

> "I'm struggling with this new business environment of ours. I'm doing the best I can, but I'm scared. A little more sensitivity and patience on your part will go a long way toward helping me cope."[5]

Leaders who listen to employees can learn a valuable lesson. It's not that they dislike change so much. Rather, it's that they don't like how it's done. The key to winning support of employees for change is involvement. Make them part of the process from the beginning. Give them a voice in how the change is implemented. Make sure that change is something done *with* employees rather than *to* them.

From the perspective of employees, managers are often viewed as the "bad guys" when changes are made. This viewpoint is just as unfair and counterproductive as the one that sees employees as inhibitors of change. This quote is typical of how managers often feel about change:

> "Maybe you see me as the instigator or 'perpetrator' of change. If you do, to a degree, you're right. Sponsoring and supporting change is one of my responsibilities—and it's an absolute necessity in order to keep our organization successful and protect our jobs. But besides being a source of change, I am also a victim of it. And when it comes to 'rolling downhill,' I end up having to make as many adjustments as anyone else."[6]

To respond effectively to change, organizations must continually apply at least the following strategies:

- Promote a "we are in this together" attitude toward positive change.
- Make sure all employees understand that change is driven by market forces, not management.
- Involve everyone who will be affected by change in planning and implementing the change.

RESTRUCTURING AND CHANGE

Few words can strike as much fear into the hearts of employees at all levels as *restructuring*. The term at one time was synonymous with *reorganization*. How-

Leadership Tip

"Business, more than any other occupation, is a continual dealing with the future; it is a continual calculation, an instinctive exercise in foresight."[7]

—Louis E. Boone

ever, because of the way so many organizations have used the word, it has become a euphemism for layoffs, terminations, closing, and workforce cuts. As a result, typical employee responses to the term are uncertainty and panic.[8]

Because of the ever-changing conditions of the global marketplace, few organizations will escape the necessity for restructuring, and few people will complete a career without experiencing one or more restructurings. Acquisitions, mergers, buyouts, and downsizing—common occurrences in today's marketplace—all typically involve corporate restructuring. This fact is market driven and can be controlled by neither individuals nor organizations. However, organizations and individuals can control how they respond to the changes brought about by restructuring, and it is this response that will determine the effectiveness of the restructuring effort. Following are some strategies for effectively leading the changes inherent in restructuring.

Be Smart and Empathetic

Restructuring can be traumatic for employees. Leaders should remember this point when planning and implementing the changes that go with restructuring. The challenge is to be both smart and empathetic. The following strategies can help maintain employee loyalty and calm employee fears during restructuring:

- Take time to show employees that management cares and is concerned about them on a personal level.
- Let employees vent.
- Communicate with employees about why the changes are necessary. Focus on market factors. Use a variety of tools to ensure effective communication (e.g., face-to-face meetings, newsletters, videotaped messages, posted notices).
- Provide formal outplacement assistance to all employees who will lose their jobs.
- Be fair, equitable, and honest with employees. Select employees to be laid off according to a definite set of criteria rather than conduct a witch-hunt.
- Remember to provide support to those individuals who will be the primary change agents.

Have a Clear Vision

One of the best ways to minimize the disruptive nature of change is to have a clear vision of what the organization is going to look like after the change. A good question to ask is "What are we trying to become?" Leaders should have a clear vision and be able to articulate that vision. This will give the organization a beacon to guide it through the emotional fog that can accompany change.

Establish Incentives That Promote the Change

People respond to incentives, especially when those incentives are important to them on a personal level. Leaders can promote the changes that accompany restructuring by establishing incentives for contributors to those changes. Incentives can be monetary or nonmonetary, but they should motivate employees on a personal level.

An effective way to identify incentives that will work is to form an ad hoc task force of employees and discuss the issue of incentives. List as many monetary and nonmonetary incentives as the group can identify. Then give the members a week to discuss the list with their fellow employees. Once a broad base of employer input has been collected, the task force meets again and ranks the incentives in order of preference. The team then establishes a menu of incentives that management can use to promote change. The menu concept allows employees to select incentives from among a list of options. This increases the likelihood that the incentives will motivate on a personal level.

Continue to Train

During times of intense change, the tendency of organizations is to put training on hold. The idea is "we'll get back to training again when things settle down." In reality, putting off training during restructuring is the last thing an organization should do.

One of the primary reasons employees sometimes oppose change is that it will require skills they don't have. Training should actually be increased during times of intense change to make sure that employees have the skills required during and after the transition period.

HOW TO LEAD CHANGE

A critical aspect of leadership in today's globally oriented organization involves leading change. To survive and thrive in a competitive environment, organizations must be able to anticipate and respond to change effectively. However, successful organizations don't just respond to change—they get out in front of it.

Leadership Profile **Charles Heimbold and Change at Bristol-Myers Squibb[9]**

Bristol-Myers Squibb is a world-leading pharmaceutical company with special expertise in the areas of treatments for cardiovascular diseases and infectious diseases, and anticancer drugs.[10] Few industries change as rapidly as those that fit under the broad health care umbrella. When Charles Heimbold took over as CEO of Bristol-Myers Squibb, he faced major market changes to which the company had to respond. He faced health care reform, patent expiration, and breast implant litigation. Any one of these would have portended major challenges and generated major changes, but all three of them together presented Heimbold with a challenge of unprecedented proportion.

Many CEOs faced with so major a challenge might have been tempted to adopt a survival strategy, but not Heimbold. Rather than trying to just hold on, he decided that Bristol-Myers Squibb would actually defy common wisdom and grow. He set a goal of doubling both sales and earnings per share. At first Heimbold's senior executives questioned the CEO's strategy. How could a company facing three potentially fatal crises even think about growing? In spite of their initial skepticism, Heimbold was determined. He persevered and led the company in achieving more than $100 million in annual global sales—an even higher level of performance than the so-called double-double that had been his original goal.[11]

Heimbold and his team at Bristol-Myers Squibb won the internal change battle, but it wasn't easy. In order to win, the company had to change not just what it did, but how it did it. Heimbold purchased promising companies, hired dynamic new managers, moved existing products into new markets, and licensed new products.[12]

One of Heimbold's greatest leadership strategies was to get all of the company's personnel to let go of the past and focus on the future. He communicated the double-double goal constantly throughout the organization so that all personnel understood what was expected, and he publicized successes to create and maintain momentum. Heimbold wanted his personnel to believe that the double-double goal not just could be achieved, but would be achieved. And it was.[13]

Because change is a constant in today's global business environment, organizations must structure themselves for it. In other words, they must institutionalize the process of change. In his book *Leading Change*, John Kotter proposes the following six-step model for effectively leading change.[14]

Educate Employees about the Need for Continual Change

Change is not something organizations do because they want to or because they get bored with the status quo. Rather, it is something they do because they must. Every organization that has to compete in an open marketplace is forced by macroeconomic conditions to constantly reduce costs, improve quality, enhance product attributes, increase productivity, and identify new markets. None of these things can be accomplished without changing the way things are currently done. Employees at all levels in an organization need to understand this point.

If it is true that change is the only constant on the radar screen of the modern organization, why do so many miss the point? A typical response is that people don't like change. But in reality this is not always the case. Employees who object to change typically object to how it is handled, not the fact that it's happening. Kotter identifies the following reasons that employees may not understand the reality of and need for change:[15]

- Absence of a major crisis
- Low overall performance standards
- No view of the big picture
- Internal evaluation measures that focus on the wrong benchmarks
- Insufficient external feedback
- A "kill the messenger" mentality among managers
- Overfocus among employees on the day-to-day stresses of the job
- Too much "happy talk" from executives

An organization's senior leadership team is responsible for eliminating any of these factors, except, of course, a major crisis caused by external forces. Creating an artificial crisis is a questionable strategy, but correcting the remainder of these factors is not just appropriate but also advisable.

Performance standards should be based on what it takes to succeed in the global marketplace. Every organization should have a strategic plan that puts the "big picture" into writing, and every employee should know the big picture. Internal evaluation measures should mirror overall performance standards in that they should ensure globally competitive performance.

Organizations cannot succeed in a competitive environment without systematically collecting, analyzing, and using external feedback. This is how an organization knows what is going on. External feedback is the most effective way to identify the need for change.

Every employee in an organization is a potential agent for positive change. Employees attend conferences, read professional journals, participate in seminars, browse the Internet, and talk to colleagues. If leaders welcome feedback from employees, they can turn them collectively into an effective mechanism for anticipating change. On the other hand, managers who "kill the messenger" will quickly extinguish this invaluable source of continual feedback.

As a rule, employees will focus most of their attention on their day-to-day duties, which is how it should be. Consequently, management must make a special effort to communicate with employees about market trends and other big-picture issues. All such communications with employees should be accurate, thorough, and honest. Managers must make sure they don't go overboard and create panic, but to sugarcoat important issues with "happy talk" is the same as deceiving employees.

Establish and Charter the Steering Committee

"Major transformations are often associated with one highly visible leader. Consider Chrysler's comeback from near bankruptcy ... and we think of Lee Iacocca. Mention Wal-Mart's ascension from small fry to industry leader, and Sam Walton comes to mind. Read about IBM's efforts to renew itself and the story centers on Lou Gerstner. After a while, one might easily conclude that the kind of leadership that is so critical to any change could come only from a single larger-than-life person."[16]

Although a visible and visionary individual can certainly be a catalyst for organizational change, the reality is that one person alone does not often change an organization.

The media like to create the image of a knight on a charging steed who single-handedly saves the company. This story makes good press, but it rarely squares with reality. Organizations that do the best job of handling change have what John Kotter calls a "guiding coalition," a team of people who are committed to the change in question and who can make it happen. Every member of the team should have the following characteristics:

- *Authority.* Members should have the authority necessary to make decisions and commit resources.
- *Expertise.* Members should have expertise that is pertinent to the change in question so that informed decisions can be made.
- *Credibility.* Members must be well respected by all stakeholders so they will be listened to and taken seriously.
- *Leadership.* Members should have the leadership qualities necessary to drive the effort. These qualities include those listed here plus influence, vision, commitment, perseverance, and persuasiveness.

Establish Antenna Mechanisms

Leading positive change is about getting out in front of it. It's about driving change rather than letting it drive you. To do this, organizations must have mechanisms for sensing trends that will generate future change. These "antenna" mechanisms can take many forms, and the more, the better. Reading professional journals, attending conferences, studying global markets, and even reading the newspaper can help identify trends that might affect an organization. So can attentive marketing representatives.

For example, a computer company that markets primarily to colleges and universities learned that two large institutions had adopted a plan to stop purchasing personal computers. Instead, they intended to require all students to purchase their own laptops. The potential for a major change in the organization's business was identified by a marketing representative in the course of a routine call on these universities. This information allowed the computer company to quickly develop and implement a plan for getting out in front of what is likely to be a nationwide trend. The company now markets laptop computers directly to students through bookstores at both universities and is promoting the idea to colleges and universities nationwide.

All employees in an organization should have their antennae tuned to the world outside and bring anything of interest to the attention of the steering committee.

Develop a Change Vision

Employees often wonder what their company will look like after a particular change has been made. The organization's vision answers this question. Kotter calls a vision a "sensible and appealing picture of the future."[17] The following scenario illustrates how having a cogent vision for positive change can help employees accept the change more readily.

Two tour groups are taking a trip together. Each group has its own tour bus and group leader. The route to the destination and stops along the way have been meticulously planned for maximum tourist value, interest, and enjoyment. All members of both groups have agreed to the itinerary and are looking forward to every stop. Unfortunately, several miles before the first scheduled attraction, the tour group leaders receive word that a chemical spill on the main highway will require a detour that will, in turn, require changes to the itinerary.

The tour group leader for group A simply acknowledges the message and tells the bus driver to take the alternate route. To the members of group A he says only that an unexpected detour has forced a change of plans. With no more information than this to go on, the members of group A are confused and quickly become unhappy.

The tour leader for group B, however, is a different sort. She acknowledges the message and asks the driver to pull over. She then says to the members of group B: "Folks, we've had a change of plans. A chemical spill on the highway up ahead has closed down the route on which most of today's attractions are located. Fortunately, this area is full of wonderful attractions. Why don't we just have lunch on me at a rustic country restaurant? It's up the road about five miles. While you folks are enjoying lunch, I'll hand out a list of new attractions I know you'll want to see. We aren't going to let a little detour spoil our fun!"

Because they could see how things would look after the change, the members of group B accepted it and were satisfied. However, the members of group A, because they were not fully informed, became increasingly frustrated and angry. As a result, they went along with the changes only reluctantly, and in several cases, begrudgingly. Group A's tour leader kept his clients in the dark. Group B's leader gave hers a vision. The five characteristics of an effective vision for change are as follows:

1. *Imaginable.* It conveys to others a picture of how things will be after the change.
2. *Desirable.* A vision that points to a better tomorrow for all stakeholders will be well received even by those who resist change.
3. *Feasible.* To be feasible, a vision must be realistic and attainable.
4. *Flexible.* An effective vision is stated in terms that are general enough to allow for initiative in responding to ever-changing conditions.
5. *Communicable.* A good vision can be explained to an outsider who has no knowledge of the business.

Communicate the Vision to All Stakeholders

People will buy into the vision only if they know about it. The vision must be communicated to all stakeholders. A good communication package will have at least the following characteristics:

■ *Simplicity.* The simpler the message, the better. Regardless of the communication formats chosen, keep the message simple and get right to it. Don't beat around the bush, lead with rationalizations, or attempt any type of linguistic subterfuge—what politicians and journalists call "spin."

■ *Repetitiveness.* Repetition is critical when communicating a new message. Messages are like flies. If a fly buzzes past your face just once and moves on, you will probably take no notice of it. But if it keeps coming back persistently and refuses to go away, before long you will take notice of it. Repetition forces employees to take notice.

■ *Multiple formats.* It is also important to use multiple formats, such as small-group meetings, large-group meetings, newsletter articles, fliers, bulletin board notices, videotaped messages, email, and a variety of other methods. A combination of visual, reading, and listening vehicles will typically be the most effective.

■ *Feedback mechanisms.* Regardless of the communication formats used, one or more feedback mechanisms must be in place. In face-to-face meetings the feedback can be spontaneous and direct. This is always the best form of feedback. However, telephone, facsimile, email, and written feedback can also be valuable.

Incorporate the Positive Change Process

Once an organization has gone through the transformational process of change, both the change itself and the process of change should be incorporated into the organization's culture. In other words, two things need to happen. First, the major change that has just occurred must be anchored in the culture so that it becomes the normal way of doing business. Second, the six-step positive change process for facilitating change must be institutionalized.

Anchoring the new change in the organizational culture is critical. If this does not occur, the organization will quickly begin to backslide and retrench. The following strategies can help an organization anchor a major change in its culture:

- *Showcase the results.* In the first place, a change must work in order to be accepted. The projected benefits of making the change should be showcased as soon as they are realized; and, of course, the sooner, the better.

- *Communicate constantly.* Do not assume that stakeholders will automatically see, understand, and appreciate the results gained by making the change. Talk about results and their corresponding benefits constantly.

- *Remove resistant employees.* If key personnel are still fighting the change after it has been made and is producing the desired results, give them the "get with it or get out" option. This approach might seem harsh, but employees at all levels are paid to move an organization forward, not to hold it back.

Institutionalizing the process of positive change is an important and final element of this step. Change is not something that happens once and then goes away. It is a constant in the lives of every person, in every organization. Consequently, the change facilitation model must become part of normal business operations. Antenna mechanisms must continue to anticipate change all the time. They feed what they see into the model, and the organization works its way through each step of the change process. The better an organization becomes at doing this, the more successful it will be at competing in the global marketplace.

Leadership Tip

"Change is the law of life. And those who look only to the past or present are certain to miss the future."[18]

—John F. Kennedy
President of the United States

LESSONS ON POSITIVE CHANGE FROM SELECTED LEADERS

Following are excerpts from the lives of several leaders that exemplify some of the positive change principles set forth in this chapter. The leaders selected for inclusion are Eleanor Roosevelt, Andrew Jackson, and Randy Tobias.

Eleanor Roosevelt[19]

"The wife and political partner of President Franklin Delano Roosevelt, Eleanor Roosevelt was, in her own right, one of the nation's great leaders of reform. Arguably, the very first person she led was her husband, encouraging and guiding his own innovative programs of social change."[20]

Eleanor Roosevelt dedicated her adult life to leading social change in the United States and the world. She led numerous social reform movements, some before her husband was elected president of the United States and many afterward. One of her most important successes in bringing about social change was passage by the United Nations of a Declaration of Human Rights. The change lesson taught by the life and career of Eleanor Roosevelt is as follows:

> To bring about positive change, a leader must be willing to swim against the current of custom, norms, and culture.

In pursuing social change, Eleanor Roosevelt often found herself swimming against the tide of social inertia. The world she had to convince to change in order to enhance the rights and welfare of women, racial minorities, and the poverty-stricken did not want to change. Worse yet, it did not want to be told by a woman that it needed to change. But Roosevelt was nothing if not determined. She held weekly press conferences, undertook lecture tours, wrote a syndicated newspaper column, and had her own radio show—things no previous first lady would have even dreamed of doing. But Roosevelt knew that in order to change the conventional status quo she would have to use unconventional tactics, and she did so with great effect.

During America's Great Depression, Eleanor Roosevelt was one of the nation's most persistent, most compassionate, and most articulate friends of the disenfranchised. As a result, she became one of the most popular first ladies in America's history. Roosevelt, as a good leader should, used her popularity to break through the social logjams that stood in the way of changes she advocated.

Andrew Jackson[21]

Andrew Jackson had earned his fame as a military leader and frontiersman well before being elected seventh president of the United States. Not all historians agree on the place of Jackson's presidency in the development of democracy in America, but there is no doubt concerning his effectiveness as a leader. Before earning fame as the stalwart commander of American forces at the Battle of Chalmette—more popularly known as the Battle of New Orleans—Jackson had made a name for himself as a delegate to the Tennessee constitutional convention (1796) and as a member of Congress from the newly formed state in 1796 and 1797. One of Jackson's most controversial but courageous political stands was in opposing President George Washington's stance toward Great Britain and its Indian allies during the American War for Independence—a stance Jackson saw as much too conciliatory.[22] The change lesson taught by the life and career of Andrew Jackson is as follows:

> To lead an organization through positive change, a leader must be strong enough to withstand the criticism, scorn, and sometime even hatred of his contemporaries.

Jackson's loathing of the "aristocratic" British and any Americans who sought to emulate Britain's aristocratic ways was legendary. Jackson was a fearless and effective military commander who led his frontiersmen against both the Red Stick Creek Indians and the British. After his ragtag army of frontiersmen and wharf rats defeated the mighty British regulars at the Battle of Chalmette (more commonly known as the Battle of New Orleans), Jackson could have entered politics and been assured of immediate success. However, he continued his career in the military, fighting against the Seminole Indians and deposing the Spanish crown in Florida. Finally, after serving as territorial governor of Florida and running unsuccessfully for the presidency against John Quincy Adams, Jackson was elected president of the United States in 1828.

Jackson wanted the government of the United States to be a democracy of the "common man." To bring this about, he first had to win the support of enough common men to get elected president. Then he had to show them that the government truly belonged to them. This he did. After his inauguration, instead of riding to the White House in a carriage as his predecessors had done, Jackson rode on horseback. Common men, he knew, could not afford fancy carriages. Rather than have the traditional aristocratic inaugural ball, he simply invited all people to stop by the executive mansion and meet their new president. So many came and the reception became so rowdy that the White House was nearly wrecked. But even this added to his reputation as the president of the common man. Perhaps the greatest change brought by Jackson and his presidency was expansion of the electorate so that the so-called common man could actually participate in government. He achieved this by eliminating property ownership as a prerequisite for voting.

Perhaps the most controversial slap in the face Jackson delivered to traditionalists was his stand against the Second Bank of the United States. Established to stabilize America's economy by tightening credit and re-centering American dollars on gold and silver, the bank made life difficult for small-business owners, farmers, and people trying to make a new start in life by moving west—in other words, common people.

Traditionalists labeled Jackson a hotheaded firebrand who would lead the United States down the path of destruction. He was hated and vilified by the wealthy, large businesses and conventional politicians who saw government as an instrument to ensure that the privileged continued to enjoy their privileges. In the nation where "all men are created equal," those who were *more equal* saw Jackson as a clear and present threat to their elevated social and economic status.

Few presidents have been simultaneously more loved and more hated during their lives than Andrew Jackson. Historians still debate the place of Jackson's presidency and the quality of his leadership. He remains as controversial to present-day historians as he was to his contemporaries. However, one thing is sure. The political involvement that many Americans take for granted today has it roots in the changes brought about by the leadership of Andrew Jackson.

Randy Tobias[23]

"As Randy Tobias met with Eli Lilly employees to explain the corporate change he had begun, he would start by talking about 'the old Lilly.' He would draw an organizational congruence model illustrating the company in terms of its strategy, work, people, structure, and culture. In change management terminology, he was describing the *current state*. Then he would talk about the remarkable changes washing over the health care industry, and he would describe 'the new Lilly' he believed would be required to meet the threats and opportunities. And he would explain every component of the new Lilly—the new strategy, the new culture and values, the new skills and attitudes that would be required of its people. To Tobias, the picture was crystal clear, and he could easily describe it to others. It was his vision of Lilly's *future state*."[24] The lesson taught by the work of Randy Tobias at Eli Lilly is as follows:

> To lead an organization through a major change in strategy, culture, values, skills, or attitudes, a leader must be able to get it through the critical transition phase.

Tobias and Lilly were at a critical point in the process of organizational change—the transition point. The company is no longer as it was, but is not yet what it will be. As Tobias learned, "In most cases, you can get people to continue managing the current state—that means doing things the same way. You can generally count on finding people to plan for the future state—in fact, many find that exciting and challenging. It's the transition state where so many organizations stumble and fall, simply because they didn't think it through in advance."[25]

The implementation or execution plan must encompass the transition period. It's like the situation people sometimes face when they sell their house in order to build a new one and then must move out of the old house before the new house is completed. This creates a period of transition that can cause difficulties in the family. The family planned well for selling the old house and equally well for moving into the new one. But nobody planned for the transition period between houses. Remembering to plan for the critical transition phase is a must for leaders of change.

Summary

1. To facilitate change in a positive way, leaders must have a clear vision and corresponding goals, exhibit a strong sense of responsibility, be effective communicators, have a high energy level, and have the will to make change happen.

2. When restructuring, organizations should show that they care, let employees vent, communicate, provide outplacement services, be honest and fair, provide support for change agents, have a clear vision, offer incentives, and train.

3. The change facilitation model contains the following steps: educate employees about the need for continual change, charter the steering committee, develop a change vision, establish antenna mechanisms, communicate the change vision, and incorporate the change process.

Key Terms and Concepts

Antenna mechanisms	Incorporate the change process
Clear vision	Organizational inertia
Communicate constantly	Remove resistant employees
Continue to train	Restructuring and change
Feasible	Smart and empathetic
Incentives	Steering committee

Review Questions

1. List the strategies leaders can use to play a positive role in facilitating change.
2. Explain what organizations must do to respond effectively to change.
3. What can organizations do to promote a positive response to restructuring?
4. Explain each step in change facilitation.

LEADERSHIP SIMULATION CASES

The following simulations are provided to generate additional thought and discussion about the principles of leadership explained in this chapter. Readers are encouraged to consider how the situations presented in these cases might apply to them and to discuss the cases with other leaders and leadership candidates.

CASE 7.1 What Is My Role in Leading Change?

The discussion had started when Dan Gilman asked a simple question: "What is my role in leading change—what should I actually do?" The trainer leading the seminar had emphasized over and over that the CEO had to be the leader of change in the organization. There were 30 CEOs participating in the seminar. They represented some of the most prominent technology companies in the country. The question Dan Gilman asked seemed to throw the trainer off balance as if he wasn't prepared for it, or, perhaps, that he had never thought about it. When the trainer's answer to the question did not seem to satisfy the CEOs, they started discussing the issue among themselves.

Discussion Questions

1. What are several strategies leaders can use to play a positive role in guiding their organizations through change?
2. What can a leader do to be a "driver" in the change process?

CASE 7.2 We Are Miles Apart on This New Change

Janice Henson's best friend is Marge Volker. It is an interesting friendship—some would say a strange friendship. Henson is vice president for information technology at ATC Technologies. Volker is a technical writer in the marketing department of ATC and the union representative for ATC technical professionals. Henson and Volker disagree on almost everything, from work issues to politics. But they also admire and respect each other. By a strange coincidence, the two women gave birth to girls in the same hospital on the same day more than 20 years ago. Their friendship began in the hospital and grew over the years as they shared the common bond and common problems of motherhood and balancing their family and work lives. Their friendship transcends work relationships and their trust in each other is complete. Today they are having lunch to discuss a major change Henson wants to make in their division. The change has generated some controversy and some hostility between management and employees. Henson asked her friend, "Marge, what is really bothering our employees about this change? How do they view the change?" Volker responded, "Good questions, Janice, and I'll give you my point of view. But I want you to do the same for me. I need to know how management views the change."

Discussion Questions

1. What might Volker tell Henson about how employees sometimes view change?
2. What might Henson tell Volker about how managers sometimes view change?
3. What strategies might Henson apply to effectively lead her organization through the current change?

CASE 7.3 I'm Worried About How Employees Will Respond to the Restructuring

"What's bothering you, John? Lately you haven't been your old positive self." John Seton is CEO of Seton Technologies, a family business that was founded by his great-grandfather. The problem he faces is this. The world in which Seton Technologies must compete has changed radically, but Seton Technologies hasn't. In some ways, the company is still operating as it did when Seton's great-grandfather ran it. "It's this restructuring I've planned," answered Seton. "Either we change and change radically or we go out of business. Those are my options. Many of our employees have been with this company since before I was born. They like the way things are now. They are not going to want to change. I'm afraid that I am going to be viewed as a traitor to the legacy of my father, grandfather, and great-grandfather."

Discussion Questions

1. Have you ever undergone a major organizational restructuring? What happened?
2. What can John Seton do to effectively lead his company through the necessary restructuring?

Endnotes

[1] David C. Shanks, "The Role of Leadership in Strategy Development," *Journal of Business Strategy* (January/February 1989): 36.

[2] Donna Deeprose, "Change: Are You a Driver, a Rider, or a Spoiler?" *Supervisory Management* (February 1990): 3.

[3] Donna Deeprose, 3.

[4] Donna Deeprose, 3.

[5] Eric Harvey and Steve Venture, *Walk Awhile in My Shoes* (Dallas: Performance Publishing, 1998), 1.

[6] Eric Harvey and Steve Venture, 1.

[7] Louis E. Boone, *Quotable Business*, 2nd ed. (New York: Random House, 1999), 303.

[8] This section is based on Alastair Rylatt, "Coping with Restructuring," in *American Management International* (May 1998): 4–5.

[9] Thomas J. Neff and James M. Citrin, *Lessons from the Top* (New York: Currency/Doubleday, 2001), 169–173.

[10] Thomas J. Neff and James M. Citrin, 172.

[11] Thomas J. Neff and James M. Citrin, 171.

[12] Thomas J. Neff and James M. Citrin, 171.

[13] Thomas J. Neff and James M. Citrin, 171.

[14] John P. Kotter, *Leading Change* (Boston: Harvard Business School Press, 1996), 3.

[15] John P. Kotter, 40.

[16] John P. Kotter, 51.

[17] John P. Kotter, 71.

[18] Louis E. Boone, 303.

[19] Alan Axelrod, *Profiles in Leadership* (Upper Saddle River, NJ: Prentice Hall, 2003), 440–443.

[20] Alan Axelrod, 440.

[21] Alan Axelrod, 245–251.

[22] Alan Axelrod, 245, 247.

[23] David A. Nadler, Champions of Change (San Francisco: Jossey-Bass, 1998), 85–87.

[24] David A. Nadler, 85.

[25] David A. Nadler, 85.

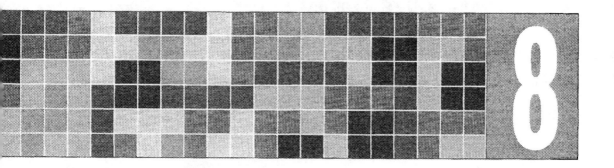

Be an Effective Team Builder

"Even the most talented individual cannot outperform a well-developed, effectively led team."

—The Author

OBJECTIVES

- Provide an overview of team building and teamwork.
- Explain the steps for building teams and making them work effectively.
- Explain the four-step approach to team building.
- Explain why teams are not bossed—they are coached.
- Demonstrate how to handle conflicts in teams.
- Explain the structural inhibitors of teamwork.
- Explain how to reward team and individual performances.
- Explain how to recognize teamwork and team players.
- Summarize the team-building lessons from selected effective leaders.

OVERVIEW OF TEAM BUILDING AND TEAMWORK

Teamwork is fundamental to success in a competitive environment for a simple and practical reason:

> Someone may be great at his or her job, maybe even the best there ever was. But what counts at work is the organization's success, not personal success. After all, if your organization fails, it does not matter how great you were; you are just as unemployed as everyone else.[1]

What Is a Team?

A team is a group of people with a common goal. The collective goal aspect of teams is critical, as is evident in the performance of athletic teams. For example, a basketball team in which one player hogs the ball, plays the role of the prima donna, and pursues his or her own personal goals (a personal-high point total, MVP status, publicity, etc.) will rarely win against a team of players all of whom pull together toward the collective goal of winning the game.

Rationale for Teams

An effective team's collective ability is greater than the sum of the abilities of its individual members. This is one of the primary reasons for advocating teamwork. In their book *TQM Team-Building and Problem-Solving*, Perry L. Johnson, Rob Kantner, and Marcia Kikora describe the rationale for teams as follows:[2]

- Two or more heads are better than one.
- The whole (the team) is greater than the sum of its parts (individual members).
- People in teams get to know each other and build trust, and as a result want to help each other.
- Teamwork promotes better communication.

It is well established that teams can outperform individuals, provided they are properly handled. A team is not just a group of people. A group of people becomes a team when the following conditions exist:

- Agreement exists as to the team's mission. For a group to be a team and a team to work effectively, all members must understand and agree on the mission.

- Members adhere to team ground rules. A team must have ground rules that establish the framework within which the team's mission is pursued. A group becomes a team when there is agreement as to mission and adherence to ground rules.

- Fair distribution of responsibility and authority exists. Teams do not eliminate structure and authority. Football teams have quarterbacks, and baseball teams have captains. However, teams work best when responsibility and authority are shared and team members are treated as equals.

- Members adapt to change. Change is not just inevitable in a competitive environment—it is desirable. Unfortunately, people often resist change. People in teams can help each other adapt to change in a positive way.

Learning to Work Together

A group of people does not a team make. People in a group do not automatically or magically find ways to work together.

One of the reasons teams don't always work as well as they might is certain built-in human factors that, unless understood and dealt with, can work against success. In *The Team Handbook* (1992), Peter R. Scholtes describes these factors as follows:[3]

Personal identity of team members.   It is natural for people to wonder where they fit into any organization. This tendency applies regardless of whether the organization is a company or a team within a company. People worry about being an outsider, getting along with other team members, having a voice, and developing mutual trust among team members. The work of the team cannot proceed effectively until team members feel as if they fit in.

Relationships among team members.   Before people in a group can work together, they have to get to know each other and form relationships. When people know each other and care about each other, they will go to great lengths to support one another. Time spent helping team members get acquainted and establish common ground is time invested well. This is especially important now that the modern workforce has become so diverse; common ground among team members can no longer be assumed.

Identity within the organization.   This factor has two aspects. The first has to do with how the team fits into the organization. Is its mission a high priority in the company? Does the team have support at the highest management levels? The second aspect relates to how membership on a given team will affect relationships with nonteam members. This concern is especially important in the case of ad hoc and project teams. Members of these teams will want to maintain relationships they have already established with fellow employees who are not on the team. They may be concerned that membership on the team might have a negative impact on their relationships with fellow workers who aren't included.

Leading Teams

Managers in the modern workplace must be able to lead teams. Mary Walsh Massop recommends the following strategies for being an effective team leader:[4]

Be clear on the team's mission.   The team's first organizational meeting should be used to draft a mission statement. This task, although guided by the team leader, should involve all team members. The statement should explain the reasons for the team's existence and define the limits of its authority. The mission statement will be the yardstick against which team performance is measured.

Identify success criteria.   The team must define what constitutes success and put it in writing. Remember, in a competitive environment success should be defined in terms of the customer—internal and external. This means that team members must understand the needs and expectations of its customers before identifying success criteria.

Be action centered.   For every success criterion, the team should develop an action statement or plan that specifies exactly what must be done to satisfy the criterion, the time frame within which it must be done, and by whom.

Establish the ground rules.   The team needs to decide how it will operate. Leaders should work to achieve consensus on such issues as these:

- Calling meetings only when necessary
- Making sure all team members come to meetings well briefed and fully prepared
- Determining how much time to allocate for agenda items
- Encouraging participants to be brief
- Determining who will serve as the recorder during meetings
- Deciding how and when to divide into subgroups
- Keeping disturbances and interruptions out of meetings
- Finishing an agenda item before moving on to the next item
- Allowing time for informed interaction among members before and after meetings

Share information.   Information should be shared freely inside and outside the team. Communication is a fundamental element of teamwork; everyone should know what is going on. Teams do not operate in a vacuum.

They are all part of a larger team, the organization. Keeping everyone informed of team activities will eliminate idle, nonproductive, and typically inaccurate speculation.

Cultivate team unity. During the 1992 Summer Olympic Games in Barcelona, Spain, the United States fielded what many sports experts and fans thought was the best basketball team ever assembled. Aptly named, the Dream Team won a gold medal with ease. But in spite of the enormous talent of individual team members (Magic Johnson, Michael Jordan, Larry Bird, Moses Malone, Scotty Pippen, and others), the gold medal was really won because these incredibly talented athletes developed a sense of unity, a team identity. As a result, they put aside individual goals, left their egos in the locker room, and played as a team—each supporting the other and contributing to the team's score rather than his own individual statistics. This does not mean they gave up their individual identities completely. On the contrary, the more team unity grew, the more supportive the team became of the individual performances.

Team Excellence and Performance

Teamwork is not a magic cure-all. Poorly run teams can do more damage to an organization's performance and corresponding competitiveness than having no teams at all. For this reason, it is critical that excellence in team performance be an overriding goal of leaders.

Dennis King, of Procter & Gamble, recommends the following strategies, which he calls the Ten Team Commandments:[5]

1. *Interdependence.* Team members should be mutually dependent on each other for information, resources, task accomplishment, and support. Interdependence is the glue that will hold a team together.
2. *Stretching tasks.* Teams need to be challenged. Responding to a challenge as a team builds team spirit and instills pride and unity.
3. *Alignment.* An aligned team is one in which all members not only share a common mission but also are willing to put aside individualism to accomplish it.
4. *Common language.* Teams often consist of members from different departments (programming, IT, project management), which typically have their own indigenous languages that may be foreign to people from other departments. Consequently, it is important for team leaders to ensure that department-specific terms and phrases are used minimally and are fully explained in common terms when employed.
5. *Trust and respect.* For team members to work well together, there must be trust and respect. Time and effort spent building trust and respect among team members is time invested well.

6. *Shared leadership and followership.*   Some group members tend to emerge as more vocal, while others sit back and observe. If group leaders are allowed to dominate, the group will not achieve its full potential. A better approach is to draw out the special talents of each individual group member so that leadership and followership are shared.

7. *Problem-solving skills.*   Time invested in helping group members become better problem solvers is time well spent. Much of the business of groups is problem solving.

8. *Confrontation- and conflict-management skills.*   Conflict is inevitable in a high-pressure, competitive workplace. Even the best teams and closest families have disagreements. Learning to disagree without being disagreeable and to air disagreement openly and frankly—attacking ideas, issues, and proposed solutions without attacking the people proposing them—are critical skills in a competitive environment.

9. *Assessment and action.*   Assessment is a matter of asking and answering the question "How are we doing?" The yardsticks for answering this question are the group's mission statement and corresponding action plan. The action plan contains goals, objectives, timetables, and assignments of responsibility. By monitoring these continually, group members can assess how the group is doing.

10. *Celebration.*   An effective team reinforces its successes by celebrating them. Recognition of a job well done can motivate team members to work even harder and smarter to achieve the next goal.

BUILDING TEAMS AND MAKING THEM WORK EFFECTIVELY

Part of building a successful team is choosing team members wisely. This section describes strategies for selecting team members, assigning responsibilities, creating a mission statement, and developing collegial relations among team members.

Leadership Tip

"Our team is well balanced. We have problems everywhere." [6]

—Tommy Prothro
Football Coach

Makeup and Size of Teams

Teams should be composed of those people who are most likely to be able to satisfy the team's mission efficiently and effectively. The appropriate makeup of a team depends in part on the type of team in question (whether it is departmental, process improvement, task force, or project oriented). Departmental teams are made up of the employees of a given department. However, process improvement teams and task forces typically cross departmental lines (cross-functional teams).

The membership of such teams should be open to any level of employee—management, supervisors, and hourly wage earners. A good rule of thumb is that the greater the mix, the better. According to Johnson et al., "The bigger the mix, the better the results. As for the size of the team, we want a group large enough to guarantee a mix of people and opinions, yet small enough to make meetings comfortable, productive, and brief. So, what's the right number? No fewer than six, no more than twelve. Eight or nine is just about right."[7]

Choosing Team Members

When putting together a team, the first step is to identify all potential team members. This is important because there will often be more potential team members than the number of members actually needed. After the list has been compiled, team members can be selected. However, care should be taken to ensure a broad mix, as discussed in the previous section. This rule should be adhered to even if there are no volunteers and team members must be drafted. The more likely case is that there will be more volunteers than openings on most teams. Johnson et al. recommend handling this by periodically rotating the membership, preferably on an alternating basis so that the team always includes both experienced and new members.[8]

Responsibilities of Team Leaders

Scholtes lists the following as responsibilities of team leaders:[9]

- Serve as the official contact between the team and the rest of the organization.
- Serve as the official record keeper for the team. Records include minutes, correspondence, agendas, and reports. Typically, the team leader will appoint a recorder to take minutes during meetings. However, the team leader is still responsible for distributing and filing minutes.

■ Serve as full-fledged team member, but with care to avoid dominating team discussions.

■ Implement team recommendations that fall within the team leader's realm of authority, and work with upper management to implement those that fall outside it.

Leadership Tip

"The team leader is the person who manages the team: calling and facilitating meetings, handling or assigning administrative details, orchestrating all team activities, and overseeing preparations for reports and presentations. The team leader should be interested in solving the problems that prompted this project and be reasonably good at working with individuals and groups. Ultimately, it is the leader's responsibility to create and maintain channels that enable team members to do their work."[10]

—Peter Scholtes
Teamwork Expert

Other Team Members

In addition to the team leader, most teams will need a team recorder and a quality adviser. The recorder is responsible for taking minutes during team meetings and assisting the team leader with the various other types of correspondence generated by the team. The quality adviser is an important part of the team in a competitive environment. Scholtes lists the following as responsibilities of the quality adviser:[11]

■ Focus on team processes rather than products and on *how* decisions are made as opposed to *what* decisions are made.

■ Assist the team leader in breaking tasks down into component parts and assigning the parts to team members.

■ Help the team leader plan and prepare for meetings.

■ Help team members learn to use the scientific approach of collecting data, analyzing data, and drawing conclusions based on the analysis.

■ Help team members convert their recommendations into presentations that can be made to upper management.

Creating the Team's Mission Statement

After a team has been formed, a team leader selected, a reporter appointed, and a quality adviser assigned, the team is ready to draft its mission statement. This is a critical step in the life of a team. The mission statement explains the team's reason for being. A mission statement is written in terms that are broad enough to encompass all it will be expected to do but specific enough that progress can be easily measured. The following sample mission statement meets both of these criteria:

> The purpose of this team is to reduce the time between when an order is taken and when it is filled, while simultaneously improving the quality of products shipped.

This statement is broad enough to encompass a wide range of activities and to give team members room within which to operate. The statement does not specify by how much throughput time will be reduced or by how much quality will be improved. The level of specificity comes in the goals set by the team (e.g., reduce throughput time by 15 percent within six months; improve the customer satisfaction rate to 100 percent within six months). Goals follow the mission statement and explain it more fully in quantifiable terms.

This sample mission statement is written in broad terms, but it is specific enough that team members know they are expected to simultaneously improve both productivity and quality. It also meets one other important criterion: simplicity. Any employee could understand this mission statement. It is brief and to the point, but comprehensive.

Team leaders should keep these criteria in mind when developing mission statements: broadness, appropriate specificity, and simplicity. A good mission statement is a tool for communicating the team's purpose—within the team and throughout the organization—not a device for confusing people or an opportunity to show off literary dexterity.

Developing Collegial Relationships

A team works most effectively when individual team members form positive, mutually supportive peer relationships. These are collegial relationships, and they can be the difference between a high-performance team and a mediocre one. Scholtes recommends the following strategies for building collegial relationships among team members:[12]

- Help team members understand the importance of honesty, reliability, and trustworthiness. Team members must trust each other and know that they can count on each other.
- Help team members develop mutual confidence in their work ability.
- Help team members understand the pressures to which other team members are subjected. It is important for team members to be supportive of peers as they deal with the stresses of the job.

These are the basics. Competence, trust, communication, and mutual support are the foundation on which effective teamwork is built. Any resource devoted to improving these factors is an investment well made.

FOUR-STEP APPROACH TO TEAM BUILDING

Effective team building is a four-step process:

1. Assess
2. Plan
3. Execute
4. Evaluate

To be a little more specific, the team-building process proceeds along the following lines: (1) assess the team's developmental needs (e.g., its strengths and weaknesses); (2) plan team-building activities based on the needs identified; (3) execute the planned team-building activities; and (4) evaluate results. The steps are spelled out further in the next sections.

Assessing Team Needs

If you were the coach of a baseball team about which you knew very little, what is the first thing you would want to do? Most coaches in such situations would begin by assessing the abilities of their new team. Can we hit? Can we pitch? Can we field? What are our weaknesses? What are our strengths? With these questions answered, the coach will know how best to proceed with team-building activities.

This same approach can be used in the workplace. A mistake commonly made by organizations is beginning team-building activities without first assessing the team's developmental needs. Because resources are limited, it is important to use them as efficiently and effectively as possible. Organizations that begin team-building activities without first assessing strengths and weaknesses run the risk of wasting resources in an attempt to strengthen characteristics that are already strong, while at the same time overlooking characteristics that are weak.

For workplace teams to be successful, they should have at least the following characteristics:

- Clear direction that is understood by all members
- "Team players" on the team
- Fully understood and accepted accountability measures

Figure 8.1 is a tool that can be used for assessing the team-building needs of (1) understanding, (2) characteristics of team members, and (3) accountability. Individual team members record their perceptions of the team's performance and abilities relative to the specific criteria in each category. The highest score possible for each criterion is 6; the lowest score possible, 0. The team score for each criterion is found by adding the scores of individual members for that criterion and dividing by the number of team members.

Instructions

To the left of each item is a blank for recording your perception regarding that item. For each item, record your perception of how well it describes your team. Is the statement completely true (CT), somewhat true (ST), somewhat false (SF), or completely false (CF)? Use the following numbers to record your perception.

> CT = 6
> ST = 4
> SF = 2
> CF = 0

Directions and Understanding

_____ 1. The team has a clearly stated mission.

_____ 2. All team members understand the mission.

_____ 3. All team members understand the scope and boundaries of the team's mission.

_____ 4. The team has a set of broad goals that support its mission.

_____ 5. All team members understand the team's goals.

_____ 6. The team has identified specific activities that must be completed to accomplish team goals.

_____ 7. All team members understand the specific activities that must be completed to accomplish team goals.

_____ 8. All team members understand projected time frames, schedules, and deadlines relating to specific activities.

Characteristics of Team Members

_____ 9. All team members are open and honest with each other at all times.

_____ 10. All team members trust each other.

_____ 11. All team members put the team's mission and goals ahead of their own personal agendas all of the time.

_____ 12. All team members are comfortable that they can depend on each other.

_____ 13. All team members are enthusiastic about accomplishing the team's mission and goals.

_____ 14. All team members are willing to take responsibility for the team's performance.

(Continues)

FIGURE 8.1   Team-building needs assessment.

_____ 15. All team members are willing to cooperate in order to get the team's mission accomplished.

_____ 16. All team members will take the initiative in moving the team toward its final destination.

_____ 17. All team members are patient with each other.

_____ 18. All team members are resourceful in finding ways to accomplish the team's mission in spite of difficulties.

_____ 19. All team members are punctual when it comes to team meetings, other team activities, and meeting deadlines.

_____ 20. All team members are tolerant and sensitive to the individual differences of team members.

_____ 21. All team members are willing to persevere when team activities become difficult.

_____ 22. The team has a mutually supportive climate.

_____ 23. All team members are comfortable expressing opinions, pointing out problems, and offering constructive criticism.

_____ 24. All team members support team decisions once they are made.

_____ 25. All team members understand how the team fits into the overall organization/big picture.

Accountability

_____ 26. All team members know how team progress/performance will be measured.

_____ 27. All team members understand how team success is defined.

_____ 28. All team members understand how ineffective team members will be dealt with.

_____ 29. All team members understand how team decisions are made.

_____ 30. All team members know their respective responsibilities.

_____ 31. All team members know the responsibilities of all other team members.

_____ 32. All team members understand their authority within the team and that of all other team members.

_____ 33. All team goals have been prioritized.

_____ 34. All specific activities relating to team goals have been assigned appropriately and given projected completion dates.

_____ 35. All team members know what to do when unforeseen inhibitors impede progress.

(Continued)

FIGURE 8.1 Team-building needs assessment.

Team-building activities should be developed and executed based on what is revealed by this assessment. Activities should be undertaken in reverse order of the assessment scores (e.g., lower scores first, higher scores last). For example, if the team score for criterion 1 (clearly stated mission) is the lowest score for all the criteria, the first team-building activity would be the development of a mission statement.

Planning Team-Building Activities

Team-building activities should be planned around the results of the needs assessment. Consider the example of a newly chartered team. The highest score for a given criterion in Figure 8.1 is 6. Consequently, any team average score less then 6 indicates a need for team building relating to the criterion in question. The lower the score, the greater the need.

For example, say the team in question had an average score of 3 for criterion 2 (all members understand the mission). Clearly, part of the process of building this team must be explaining the team's mission more clearly. A team average score of 3 on this issue indicates that some members understand the mission and some don't. The solution might be as simple as having the team leader sit down with the team, describe the mission, and respond to questions from team members.

On the other hand, if the assessment produces a low score for criterion 9 (all team members are open and honest with each other all the time), more extensive trust-building activities may be needed. In any case, what is important in this step is to (1) plan team-building activities based on what is learned from the needs assessment and (2) provide team-building activities in the priority indicated by the needs assessment, beginning with the lowest scores.

Executing Team-Building Activities

Team-building activities should be implemented on a "just-in-time" basis. A mistake made by many organizations is rushing into team building, giving all employees teamwork training, even those who are not yet part of a chartered team. Like any kind of training, teamwork training will be forgotten unless it is put to immediate use. Consequently, the best time to provide teamwork training is after a team has been formed and given its charter. In this way, team members will have opportunities to apply what they are learning immediately.

Team building is an ongoing process. The idea is to make a team better and better as time goes by. Consequently, basic teamwork training is provided as soon as a team is chartered. All subsequent team-building activities are based on the results of the needs-assessment and planning process.

Evaluating Team-Building Activities

If team-building activities have been effective, weaknesses pointed out by the needs-assessment process should have been strengthened. A simple way to evaluate the effectiveness of team-building activities is to readminister the appropriate portion of the needs-assessment document. The best approach is to reconstitute the document so that it contains the relevant criteria only. This will focus the attention of team members on the specific targeted areas.

If the evaluation shows that sufficient progress has been made, nothing more is required. If not, additional team-building activities are needed. If a given team-building activity appears to have been ineffective, get the team together and discuss it. Use the feedback from team members to identify weaknesses and problems and use the information to ensure that team-building activities become more effective.

TEAMS SHOULD NOT BE BOSSED—THEY SHOULD BE COACHED

If employees are going to be expected to work together as a team, leaders have to realize that teams are not bossed—they are coached. Team leaders, regardless of their title (manager, supervisor, etc.) need to understand the difference between bossing and coaching. Bossing, in the traditional sense, involves planning work, giving orders, monitoring programs, and evaluating performance. Bosses approach the job from an "I'm in charge—do as you are told" perspective.

Coaches, on the other hand, are facilitators of team development and continually improved performance. They approach the job from the perspective of leading the team in such a way that it achieves peak performance levels on a consistent basis. This philosophy is translated into everyday behavior as follows:

- Coaches give their teams a clearly defined charter.
- Coaches make team development and team building a constant activity.
- Coaches are mentors.
- Coaches promote mutual respect between themselves and team members and among team members.
- Coaches make human diversity within a team a plus.

Clearly Defined Mission Statement

One can imagine a basketball, soccer, or track coach calling her team together and saying, "This year we have one overriding purpose—to win the championship." In one simple statement this coach has clearly and succinctly defined the team's mission statement. All team members now know that everything they do this season should be directed at winning the championship. The coach didn't say that the team would improve its record by 25 points, improve its standing in the league by two places, or make the playoffs, all of which would be worthy missions. This coach has a greater vision—this year the team is going for the championship. Coaches of work teams should be just as specific in explaining the team's mission to team members.

Team Development and Team Building

The most constant presence in an athlete's life is practice. Regardless of the sport, athletic teams practice constantly. During practice, coaches work on developing the skills of individual team members and the team as a whole. Team development and team-building activities are ongoing forever. Coaches of work teams should follow the lead of their athletic counterparts. Developing the skills of individual team members and building the team as a whole should be a normal part of the job—a part that takes place regularly and never stops.

Mentoring

Good coaches are mentors, establishing a helping, caring, nurturing relationship with team members. Developing the capabilities of team members, improving the contribution individuals make to the team, and helping team members advance their careers are all mentoring activities. According to Gordon F. Shea, author of *Mentoring* (1994), effective mentors help team members in the following ways:[13]

- Developing their job-related competence
- Building character
- Teaching them the corporate culture
- Teaching them how to get things done in the organization
- Helping them understand other people and their viewpoints
- Teaching them how to behave in unfamiliar settings or circumstances
- Giving them insight into differences among people
- Helping them develop success-oriented values

Mutual Respect

It is important for team members to respect their coach, for the coach to respect her team members, and for team members to respect each other. According to Robert H. Rosen (*The Healthy Company*, 1991), "Respect is composed of a number of elements that, like a chemical mixture, interact and bond together."[14]

- *Trust made tangible.* Trust is built by (a) setting the example, (b) sharing information, (c) explaining personal motives, (d) avoiding both personal criticisms and personal favors, (e) handing out sincere rewards and recognition, and (f) being consistent in disciplining.

- *Appreciation of people as assets.* Appreciation for people is shown by (a) respecting their thoughts, feelings, values, and fears, (b) respecting their desire to lead and follow, (c) respecting their individual strengths and differences, (d) respecting their desire to be involved and to participate, (e) respecting their need to be winners, (f) respecting their need to learn, grow, and develop, (g) respecting their need for a safe and healthy workplace that is conducive to peak performance, and (h) respecting their personal and family lives.

- *Communication that is clear and candid.* Communication can be made clear and candid if coaches will do the following: (a) open their eyes and ears—observe and listen, (b) say what you want, say what you mean (be tactfully candid), (c) give feedback constantly and encourage team members to follow suit, and (d) face conflict within the team head-on; that is, don't let resentment among team members simmer until it boils over—handle it now.

- *Ethics that are unequivocal.* Ethics can be made unequivocal by (a) working with the team to develop a code of ethics, (b) identifying ethical conflicts or potential conflicts as early as possible, (c) rewarding ethical behavior, (d) disciplining unethical behavior consistently, and (e) making new team members aware of the team's code of ethics before bringing them in. In addition to these strategies, the coach should set a consistent example of unequivocal ethical behavior.

Human Diversity

Human diversity is a plus. Sports and the military have typically led American society in the drive for diversity, and both have benefited immensely. To list the contributions to either sports or the military made by people of different genders, races, religions, and so on would be a gargantuan task. Fortunately, leading organizations in the United States have followed the positive example set by sports and the military. The smart ones have learned that most of the growth in the workplace will be among women, minorities, and immigrants, who will bring new ideas and varied perspectives, precisely what an organization needs to stay on the razor's edge of competitiveness. However, in spite of steps already taken toward making the

American workplace both diverse and harmonious, wise coaches understand that people—consciously and unconsciously—tend to erect barriers between themselves and people who are different from them. This tendency can quickly undermine that trust and cohesiveness on which teamwork is built. To keep this from happening, coaches can do the following:

■ *Conduct a cultural audit.* Identify the demographics, personal characteristics, cultural values, and individual differences among team members.

■ *Identify the specific needs of different groups.* Ask women, ethnic minorities, and older workers to describe the unique inhibitors they face. Make sure all team members understand these barriers, then work together as a team to eliminate, overcome, or accommodate them.

■ *Confront cultural clashes.* Wise coaches meet conflict among team members head-on and immediately. This approach is particularly important when the conflict is based on diversity issues. Conflicts that grow out of religious, cultural, ethnic, age, or gender issues are more potentially volatile than everyday disagreements over work-related concerns. Consequently, conflict that is based on or aggravated by human difference should be confronted promptly. Few things will polarize a team faster than diversity-related disagreements that are allowed to fester and grow.

■ *Eliminate institutionalized bias.* A company workforce that had historically been predominantly male now has a majority of women. However, the physical facility still has 10 men's rest rooms and only 2 for women. This imbalance is an example of institutionalized bias. Teams may unintentionally slight some members, simply out of habit or tradition. This is the concept of "discrimination by inertia." It happens when the demographics of a team changes but its habits, traditions, procedures, and work environment do not.

An effective way to eliminate institutional bias is to circulate a blank notebook and ask team members to record—without attribution—instances and examples of institutional bias. After the initial circulation, repeat the process periodically. The coach can use the input collected to help eliminate institutionalized bias. By collecting input directly from team members and acting on it promptly, coaches can ensure that discrimination by inertia is not creating or perpetuating quiet but debilitating resentment.

HANDLING CONFLICT IN TEAMS

Conflict occurs in even the best teams. Even when all team members agree on a goal, they may still disagree on how best to accomplish it. James Lucas (*Fatal Illusions*, 1996) recommends the following strategies for preventing and resolving team conflict:[15]

Leadership Profile   **Egon Zehnder Rewards Teams and Individuals at EZI**[16]

"Egon Zehnder, founder of Egon Zehnder International (EZI), Europe's largest global search firm, knows very well that if you want people to cooperate you have to reward joint effort. To him the unique compensation system at EZI is in the best interest of the clients, the firm, its consultants, and its partners. The system is quite simple, actually. After people make partner, they each get a base salary and an equal share of the profit of the firm. That's right: *an equal share.* Partners also receive another set of profit shares based solely on the length of time they have been a partner. This means that even if a five-year partner breaks records and bills twice what a ten-year partner bills, the ten-year partner will make more money than the five-year partner."[17]

Zehnder has found a way to reward collective corporate performance while still rewarding individual performance and while also acknowledging (financially) length of service to the company. By tying compensation to collective corporate performance, Zehnder promotes collaboration and teamwork among the partners. By rewarding individual performance on a base salary and the potential for individual incentives that factor in length of service, Zehnder promotes individual effort.

"People are more likely to cooperate, whether in the classroom or the corporation, if their joint efforts are rewarding. Yet growing up in a culture that rewards individualistic or competitive achievement leaves many with the perception that they'll do better if they are each rewarded solely based on their individual accomplishments. They're wrong. The fact is that cooperation pays bigger bonuses."[18]

- Plan and work to establish a culture where individuality and dissent are in balance with teamwork and cooperation.
- Establish clear criteria for deciding when decisions will be made by individuals and when they will be made by teams.
- Don't allow individuals to build personal empires or to use the organization to advance personal agendas.
- Encourage and recognize individual risk-taking behavior that breaks the organization out of unhelpful habits and negative mental frameworks.
- Encourage healthy, productive competition and discourage unhealthy, counterproductive competition.
- Recognize how difficult it can be to ensure effective cooperation, and spend the energy necessary to get just the right amount of it.
- Value constructive dissent, and encourage it.
- Assign people of widely differing perspectives to every team or problem.
- Reward and recognize both dissent and teamwork when they solve problems.

- Reevaluate the project, problem, or idea when no dissent or doubt is expressed.
- Avoid hiring people who think they don't need help or who don't value cooperation.
- Ingrain into new employees the need for balance between the concepts of cooperation and constructive dissent.
- Provide ways for employees to say what no one wants to hear.
- Realistically and regularly assess the ability and willingness of employees to cooperate effectively.
- Understand that some employees are going to clash, so determine where this is happening and remix rather than waste precious organizational energy trying to get people to like each other.
- Ensure that the organization's value system and reward and recognition systems are geared toward cooperation with constructive dissent rather than dog-eat-dog competition or cooperating at all costs.
- Teach employees how to manage both dissent (not let it get out of hand) and agreement.
- Quickly assess whether conflict is healthy or destructive, and take immediate steps to encourage the former and resolve or eliminate the latter.

Leadership Tip

"The homerun record? I don't even think about it. If it happens, it happens, but breaking records isn't what makes a team."[19]

—Ken Griffey
Baseball Player

STRUCTURAL INHIBITORS OF TEAMWORK

One of the primary and most common reasons that teamwork never gains a foothold in certain organizations is that those organizations fail to remove built-in structural inhibitors. A structural inhibitor is an administrative procedure, organizational principle, or cultural element that works against a given change—in this case, the change from individual work to teamwork. Organizations often make the mistake of espousing teamwork without first removing the structural inhibitors that will guarantee its failure.

In an article for *Quality Digest* (June 1996), Michael Donovan describes some of the structural inhibitors to effective teamwork that are commonly found in organizations:[20]

■ *Unit structure.* Teams work best in a cross-functional environment as opposed to the traditional functional-unit environment. This allows teams to be process or product oriented. Failing to change the traditional unit structure can inhibit teamwork.

■ *Accountability.* In a traditional organization, employees feel accountable to management. This perception can undermine teamwork. Teams work best when they feel accountable to customers. Managers in a team setting should view themselves as internal emissaries for customers.

■ *Unit goals.* Traditional organizations are task oriented, and their unit goals reflect this orientation. A task orientation can undermine teamwork. Teams work best when they focus on overall process effectiveness rather than individual tasks.

■ *Responsibility.* In a traditional organization, employees are responsible for their individual performance. This individual orientation can be a powerful inhibitor to teamwork. Teams work best when individual employees are responsible for the performance of their team.

■ *Compensation and recognition.* The two most common stumbling blocks to teamwork are compensation and recognition. Traditional organizations recognize individual achievement and compensate on the basis of either time or individual merit. Teams work best when both team and individual achievements are recognized and when both individual and team performances are compensated.

■ *Planning and control.* In a traditional organization, managers and supervisors plan and control the work. Teams work best in a setting in which managers and teams work together to plan and control work.

Leaders who are serious about teamwork and need the improved productivity that can result from it must begin by removing structural inhibitors. In addition to the inhibitors described earlier, leaders should be diligent in rooting out others that exist in their organizations. An effective way to identify structural inhibitors in an organization is to form employee focus groups and ask the following question: *"What existing administrative procedures, organizational principles, or cultural factors will keep us from working effectively in teams?"* Employees are closer to the most likely inhibitors daily and can, therefore, provide invaluable insight in identifying them.

REWARDING TEAM AND INDIVIDUAL PERFORMANCES

Attempts to institutionalize teamwork will fail unless they include an appropriate compensation system; in other words, if you want teamwork to work, make it pay. This does not mean that employees are no longer compensated as individuals. Rather, the most successful compensation systems combine both individual and team pay.

This matter is important because few employees work exclusively in teams. A typical employee, even in the most team-oriented organization, also spends some time involved in individual activities. Even those who work full time in teams have individual responsibilities that are carried out on behalf of the team.

Consequently, the most successful compensation systems have the following components: (1) base individual compensation, (2) individual incentives, and (3) team-based incentives. With such a system, all employees receive their traditional individual base pay. Then there are incentives that allow employees to increase their income by surpassing goals set for their individual performance. Finally, other incentives are based on team performance. In some cases the amount of team compensation awarded to individual team members is based on their individual performance within the team, or, in other words, on the contribution they made to the team's performance.

An example of this approach can be found in the world of professional sports. All baseball players in both the National and American Leagues receive a base amount of individual compensation. Most also have a number of incentive clauses in their contracts to promote better individual performance. Team-based incentives are offered if the team wins the World Series or the league championship. When this happens, the players on the team divide the incentive dollars into shares. Every member of the team receives a certain number of shares based on his perceived contribution to the team's success that year.

Figure 8.2 is a model that can be used to establish a compensation system that reinforces both team and individual performances. The first step in this model involves deciding what performance outcomes will be measured (individual and team outcomes). Step 2 involves determining how the outcomes will be measured (What types of data will tell the story? How can these data be collected? How frequently will the performance measurements be made?).

Step 3 involves deciding what types of rewards will be offered (monetary, nonmonetary, or a combination of the two). In this step rewards are pegged to levels of performance so that the reward is in proportion to the performance.

The issue of proportionality in a compensation system is important when designing incentives. If just barely exceeding a performance goal results in the same reward given for substantially exceeding it, "just barely" is what the organization will get. If exceeding a goal by 10 percent results in a 10 percent bonus, then exceeding it by 20 percent should result in a 20 percent bonus, and so on. Proportionality and fairness are characteristics that employees scrutinize with care when examining an incentive formula. Any formula that is perceived as unfair or disproportionate will not have the desired result.

The final step in the model in Figure 8.2 involves integrating the compensation system with other performance-related processes. These systems include performance appraisal, the promotion process, and staffing. If

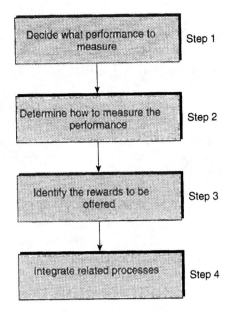

FIGURE 8.2   Model for developing a team and individual compensation system.

teamwork is important, one or more criteria relating to teamwork should be included in the organization's performance-appraisal process.

Correspondingly, an employee's ratings on the teamwork criteria in a performance appraisal should be considered when making promotion decisions. An ineffective team player should not be promoted in an organization that values teamwork. Other employees will know, and teamwork will be undermined. Finally, during the selection process, applicants should be questioned concerning their views on teamwork. It makes no sense for an organization that values teamwork to hire new employees who, during their interview, show no interest in or aptitude for teamwork.

A common mistake leaders make when organizations first attempt to develop incentives is thinking that employees will respond only to dollars in a paycheck. In fact, according to Douglas G. Shaw and Craig Eric Schneier, in an article in *Human Resource Planning*, nonmonetary rewards can be even more effective than actual dollars.[21] Widely used nonmonetary rewards that have proved to be effective include movie tickets, gift certificates, time off, event tickets, free attendance at seminars, getaway weekends for two, airline tickets, and prizes such as electronic or household products.

Different people respond to different incentives. Consequently, what will work can be difficult to predict. A good rule of thumb to apply when selecting nonmonetary incentives is "Don't assume—ask." Employees know what appeals to them. Before investing in nonmonetary incentives, organizations should survey their employees. List as many different potential nonmonetary rewards as possible and let employees rate them. In addition, set up the incentive system so that employees, to the extent possible, are able to select the reward that appeals to them. For example, employees who exceed performance goals (team or individual) by 10 percent should be allowed to select from among several equally priced rewards on the "10 Percent Menu." Whereas one employee might enjoy dinner tickets for two, another might be more motivated by tickets to a sporting event. The better an incentive program is able to respond to individual preferences, the better it will work.

RECOGNIZING TEAMWORK AND TEAM PLAYERS

One of the strongest human motivators is recognition. People don't just want to be recognized for their contributions; they *need* to be recognized. The military applies this fact very effectively. The entire system of military commendations and decorations (medals) is based on the positive human response to recognition. No amount of pay could compel a young soldier to perform the acts of bravery that are commonplace in the history of the U.S. military. But the recognition of a grateful nation continues to spur men and women on to incredible acts of valor every time our country is involved in an armed conflict. There is a lesson here for nonmilitary organizations.

The list of methods for recognizing employees goes on ad infinitum. There is no end to the ways that the intangible concept of employee appreciation can be expressed. For example, writing in *Quality Digest*, Bob Nelson recommends the following methods:[22]

- Write a letter to the employee's family members telling them about the excellent job the employee is doing.
- Arrange for a senior-level manager to have lunch with the employee.
- Have the CEO of the organization call the employee personally (or stop by in person) to say, "Thanks for a job well done."
- Find out what the employee's hobby is and publicly award him or her a gift relating to that hobby.
- Designate the best parking space in the lot for the "Employee of the Month."
- Create a "Wall of Fame" to honor outstanding performance.

These examples are provided to trigger ideas, but are only examples of the many ways that you can recognize employees. Every organization should develop its own locally tailored recognition options. When doing so, the following rules of thumb will be helpful:

- Involve employees in identifying the types of recognition activities to be used. Employees are the best judges of what will motivate them.

- Change the list of recognition activities periodically. The same activities used over and over for too long will become stale.
- Have a variety of recognition options for each level of performance. This will allow employees to select the type of reward that appeals to them most.

LESSONS ON TEAM BUILDING FROM SELECTED LEADERS

Following are excerpts from the lives of several leaders that exemplify some of the team-building principles set forth in this chapter. The leaders selected for inclusion are Shawnee Chief Tecumseh, H. Norman Schwarzkopf, and Martha Ingram.

Tecumseh[23]

Because of his credibility and charisma, the Shawnee Chief Tecumseh was able to do what no other Indian leader could—unite several warring Indian tribes against white settlement during the War of 1812, the War for American Independence from Great Britain, and the Indian Wars of the Northwest. Tecumseh had been adopted as a child by Blackfish, a prominent Shawnee Chief, who agreed to take him in after Tecumseh's father was killed in battle. Blackfish raised the fatherless boy as his own son and instilled in him the qualities that later would make Tecumseh famous. Under Blackfish's dedicated mentoring and disciplined supervision, Tecumseh became not only a courageous and skilled warrior, but a talented diplomat.[24]

Tecumseh was determined to arrest the steady erosion of what he thought of as Indian lands by the persistent push west of white settlement. But he knew that in order to have any hope of stopping the settlers, he would need to form a number of disparate Indian tribes into an alliance and join that alliance with the British. This he did, and with great effect in both the Revolutionary War and the War of 1812. The significance of his team-building effort cannot be overstated. To get the various Indian tribes to even talk without attacking each other was a great feat of diplomacy, persuasion, and leadership. While it is true that the various Indian tribes hated the white settlers, it is equally true that they hated each other. But Tecumseh knew that the previously warring Indian tribes feared white settlement of "their" lands as much as he did. He used this knowledge effectively to bring the tribes together and unite them against a common enemy. The team-building lessons of the life of Tecumseh are as follows:

> Leaders must be able to create consensus around a common goal if they want to build effective teams.

> In order to get a diverse group of people to work together as a team, leaders must have patience, endurance, and persuasion.

Tecumseh was ultimately defeated in the American Revolutionary War by General Anthony Wayne at the Battle of Fallen Timbers in August 1794. Tecumseh remained at peace for 18 years, but he never accepted white settlement of territories previously claimed by various Indian tribes. Consequently, with the outbreak of the War of 1812, Tecumseh and his "team" of Indian tribes once again joined the British. This great chief commanded a force consisting of Indians from disparate tribes as well as British troops. Through leadership, persuasion, and sheer determination, he molded his diverse force into an effective fighting team. As a result of his effectiveness in building and leading a fighting force of Indian tribes that had nothing in common except their hatred for white settlers, Tecumseh was commissioned as a brigadier general in the British army.

H. Norman Schwarzkopf[25]

"During the Persian Gulf War of 1990–91, H. Norman Schwarzkopf was one of the best known men in America. He was the overall commander of the U.S. led coalition assembled against the Iraq of Saddam Hussein."[26] The coalition led by Schwarzkopf was a disparate group of countries—each with its own agenda. Taking a loose coalition of international partners that could fall apart at any moment due to global political and economic concerns and pulling it together as a unified fighting force was a team-building exercise of immense proportions. But General Schwarzkopf did it, and did it well. The team-building lessons of H. Norman Schwarzkopf are as follows:

> To build a team and keep it functioning effectively as a team, leaders must communicate often and well.

> To hold a team together and keep it focused on its mission, the leader must demonstrate unquestionable integrity.

In August 1990 Operation Desert Shield was launched in response to Iraq's invasion of the tiny neighboring country of Kuwait. Schwarzkopf was named overall commander of a coalition force from 48 countries. When Saddam Hussein refused to withdraw from Kuwait, Desert Shield became Desert Storm and General Schwarzkopf found himself leading his diverse coalition into battle.

Not only did General Schwarzkopf have to lead a huge multinational army, he also had to overcome America's past—a past colored by fearful memories of the political debacle that turned Vietnam into a conflict in which the battles were all won, but the war was lost. Schwarzkopf not only won on the battlefield, he won over the minds of Americans by erasing the lingering doubts about "another Vietnam," thereby restoring the country's dignity and self-image. During the 100-hour ground campaign, coalition forces inflicted more than 150,000 casualties on the Iraqi army while sustaining fewer than 500 themselves.[27]

Martha Ingram[28]

Ingram Industries is one of the leading companies in the United States in the distribution, barge operation, and insurance businesses. It is also one of the largest companies in the United States owned and personally led by a woman. Ingram Book Groups leads the nation in the wholesale book distribution business. Annual shipments exceed 115 million books, audiotapes, and CDs. Ingram Marine Group operates more than 2,500 barges. Ingram Permanent General Insurance is a leader in providing coverage for high-risk drivers.[29]

Martha Ingram had been working alongside her husband, Bronson, the CEO of Ingram Industries for more than 16 years at the time of his death. She never really wanted to be CEO of the company, but her husband—aware of her outstanding organizational skills honed from years of running Ingram Industry's corporate donations program—asked Martha to join the executive management team in case anything should happen to him. The Ingrams had four children. Prior to joining the company's management team, Martha Ingram had concentrated on being a mother and a performing arts patron. She helped establish the Nashville Ballet, the Tennessee Performing Arts Center, and the Tennessee Repertory Theater in addition to helping improve the Nashville Symphony Orchestra. Martha Ingram also raised her boys and hoped to see them follow in their father's footsteps.

The Ingram children were all involved in the operation of the business when Bronson Ingram died. They were being honed to someday take over the company their father had built, but upon his death the sons were concerned that their youth might be viewed with skepticism should one of them assume the CEO position. They "drafted" their mother to assume the leadership void created by Bronson's death. As it turned out, this decision was prophetic. Not only did Martha Ingram keep the company operating smoothly, she actually expanded its scope and its profits. She also placed the operation of each of the company's various divisions under the leadership of one of her sons. Each son had a distinct division of the company to run.

One of the ways that Martha Ingram improved the performance of a company that was already performing very well was to apply her motherly instincts in building the company's entire management team and workforce—not just her sons—into a "family". The team-building lesson of Martha Ingram is as follows:

Teams function better when individual team members view each other as family.

Martha Ingram knew what it took to maintain a family, and she applied that knowledge in transforming the workforce of Ingram Industries into a family. Because communication is so important in a family, Ingram installed a toll-free line employees could use to talk directly to her if their problems could not be worked out through normal channels. The "hotline" rings only in Ingram's office. This strategy not only succeeded in bringing Ingram and her far-flung workforce closer together, it had the added benefit of keeping her in touch with a company that is spread out all over the globe. Ingram credits her success as a CEO to her experience in being a mother. Both jobs involve keeping people feeling good about themselves and the family and doing their best to promote the best interests of both.[30]

Summary

1. A team is a group of people with a common goal. The rationale for the team approach to work is that "two heads are better than one." A group of people becomes a team when the following conditions exist: there is agreement as to the mission, members adhere to ground rules, there is a fair distribution of responsibility and authority, and people adapt to change.

2. To be an effective team leader, one should apply the following strategies: be clear on the team's mission, identify success criteria, be action centered, establish ground rules, share information, and cultivate team unity. One can be a good team member by applying the following strategies: gain entry, be clear on the team's mission, be well prepared and participate, and stay in touch.

3. The Ten Team Commandments are interdependence, stretching tasks, alignment, common language, trust and respect, shared leadership and followership, problem-solving skills, confrontation- and conflict-handling skills, assessment and action, and celebration.

4. After a team has been formed, a mission statement should be drafted. A good mission statement summarizes the team's reason for being. It should be broad enough to encompass all the team is expected to do but specific enough to allow for the measurement of progress.

5. Teams are not bossed—they are coached. Coaches are facilitators and mentors. They promote mutual respect among team members and foster cultural diversity.

6. Employees will not always work well together as a team just because it's the right thing to do. Employees might not be willing to trust their performance, in part, to other employees.

7. Common structural inhibitors in organizations are unit structure, accountability, unit goals, responsibility, compensation, recognition, planning, and control.

8. Team and individual compensation systems can be developed in four steps: (1) decide what performance to measure, (2) determine how to measure the performance, (3) identify the rewards to be offered, and (4) integrate related processes.

Key Terms and Concepts

Alignment	Mission statement
Celebration	Needs assessment
Coaching	Nonmonetary rewards
Collegial relationships	Proportionality
Common language	Structural inhibitor
Conflict	Team building
Diversity	Teamwork
Interdependence	Trust building

Review Questions

1. What is a team and why are teams important?
2. When does a group of people become a team?
3. Explain the strategies for being an effective team leader.
4. List and explain the Ten Team Commandments.
5. What are the characteristics of a good team mission statement?
6. Explain the concept of collegial relationships.
7. Describe how to promote diversity in teams.
8. Explain the concept of institutionalized bias.
9. Explain why some employees are not comfortable being team players.
10. List and describe four common structural inhibitors of teamwork in organizations.
11. Explain the concept of nonmonetary rewards.

LEADERSHIP SIMULATION CASES

The following simulations are provided to generate additional thought and dis-cussion about the principles of leadership explained in this chapter. Readers are encouraged to consider how the situations presented in these cases might apply to them and to discuss the cases with other leaders and leadership candidates.

CASE 8.1 Stop Bossing and Start Coaching

"I'm just not getting the performance out of my team that I need," sighed a frustrated Jan Bulland. "What do you mean?" asked Jerry Powers, Bulland's colleague at NTI Corporation. Bulland went on to explain that her team members performed well enough, although not at peak levels, when she was present and providing hands-on supervision. But any time she had to be out of the office, the team's performance slipped noticeably. Powers asked Bul-land several questions about her leadership style. After hearing her answers, he said, "Jan, you need to stop bossing and start coaching." "What do you mean?" asked Bulland. "What's the difference?"

Discussion Questions

1. Have you ever worked on a team that was bossed instead of coached? How did the team leader's approach affect the team morale and long-term performance?

2. If you were in Power's place, how would you explain the difference between bossing and coaching a team?

CASE 8.2 All This Team Ever Does Is Squabble

John Denny had just about had it with his team. To him it seemed that all his team members ever did was squabble, bicker, and fight. If the situation didn't change, heads were going to roll, or he was going to quit. But one way or another, his team members were going to get focused on their duties and stop their constant fighting.

He gave himself a weekend to cool off. Then, first thing Monday morn-ing, Denny called a team meeting. This is what he told his team members: "I am fed up with the constant bickering, backbiting, sniping, and fighting that goes on in this team. As far as I'm concerned, every one of you needs to grow up and start acting like an adult. I'm warning you. The fighting stops today!"

Discussion Questions

1. Assess Denny's approach to resolving the conflict in his team. Do you think his talk with his team will do any good in the long term?

2. Assume you are a colleague of Denny's and he asks you for advice on how to prevent and resolve conflict in his team. What would you tell him?

CASE 8.3 Teamwork Isn't Working. What's Wrong?

Juan Ortega didn't understand what was going on with his company. When he originally founded Ortega Industries, the company was small enough that everyone worked together like family. But that was years ago. Now Ortega Industries had grown to a medium-size company employing 900 people. In an attempt to recapture the family orientation of the old days, Juan Ortega had instituted a companywide program to implement teamwork.

Ortega knew the value of teamwork firsthand. In the old days, employees pulled together and provided each other with mutual support. When the company had only 50 employees, every one of them approached his job as if he owned Ortega Industries. That's the attitude Juan Ortega is trying to instill now through his teamwork implementation. The problem is, it's not working. In spite of his best efforts, it seems that most employees continue to do things the same old way they have always done them.

In an effort to understand why his teamwork implementation has not brought the desired results, Ortega has been reading every teamwork book he can find. Just last night he read about "structural inhibitors" to teamwork in one of his books. "I wonder if this is my problem?" thought Ortega.

Discussion Questions

1. How can Juan Ortega determine if there are structural inhibitors at work holding back his teamwork effort?

2. If Juan Ortega determines that structural inhibitors are holding his company back, how can he go about eliminating them?

Endnotes

1 Perry L. Johnson, Rob Kantner, and Marcia A. Kikora, TQM Team-Building and Problem-Solving (Southfield, MI: Perry Johnson, 1990), Chapter 1, p. 1.

[2] Perry Johnson, Rob Kantner, and Marcia Kikora, Chapter 1, p. 2.

[3] Peter R. Scholtes, *The Team Handbook* (Madison, WI: Joiner Associates, 1992), Chapter 6, p. 36.

[4] Mary Walsh Massop, "Total Teamwork: How to Be a Member," in *Management for the 90s: A Special Report from* Supervisory Management (New York: AMACOM, 1991), 8.

[5] Dennis King, "Team Excellence," in *Management for the 90s: A Special Report from* Supervisory Management (New York: AMACOM, 1991), 16–17.

[6] Louis E. Boone, *Quotable Business*, 2nd ed. (New York: Random House, 1999), 86.

[7] Perry Johnson, Rob Kantner, and Marcia Kikora, Chapter 2, p. 1.

[8] Perry Johnson, Rob Kantner, and Marcia Kikora, Chapter 2, pp. 2–3.

[9] Peter R. Scholtes, Chapter 3, pp. 9–10.

[10] Peter R. Scholtes, Chapter 3, p. 8.

[11] Peter R. Scholtes, Chapter 3, p. 8.

[12] Peter R. Scholtes, Chapter 3, p. 8.

[13] Gordon F. Shea, *Mentoring* (New York: American Management Association, 1994), 49–50.

[14] Robert H. Rosen, *The Healthy Company* (New York: Putnam, Perigee Books, 1991), 24.

[15] James R. Lucas, *Fatal Illusions* (New York: AMACOM, 1996), 160–161.

[16] James M. Kouzes and Barry Z. Posner, *The Leadership Challenge*, 3rd ed. (San Francisco: Jossey-Bass, 2002), 256–257.

[17] James M. Kouzes and Barry Z. Posner, 256.

[18] James M. Kouzes and Barry Z. Posner, 257.

[19] Louis E. Boone, 87.

[20] Michael Donovan, "Maximizing the Bottom Line Impact of Self-Directed Work Teams," *Quality Digest* 16, no. 6 (June 1996): 38.

[21] Douglas G. Shaw and Craig Eric Schneier, "Team Measurements and Rewards: How Some Companies Are Getting It Right," *Human Resource Planning* 18, no. 3 (1995): 39.

[22] Bob Nelson, "Secrets of Successful Employee Recognition," *Quality Digest* 16, no. 8 (August 1996): 29.

[23] Alan Axelrod, *Profiles in Leadership* (Upper Saddle River, NJ: Prentice Hall, 2003), 511–513.

[24] Alan Axelrod, 511.

[25] Alan Axelrod, 472–474.

[26] Alan Axelrod, 472.

[27] Alan Axelrod, 473.

[28] Thomas J. Neff and James M. Citrin, *Lessons from the Top* (New York: Currency/Doubleday, 2001), 175–180.

[29] Thomas J. Neff and James M. Citrin, 179.

[30] Thomas J. Neff and James M. Citrin, 178.

Empower Followers to Lead Themselves

"How well employees perform in the absence of their leader is the ultimate test of leadership."

—The Author

OBJECTIVES

- Define employee self-leadership.
- Define employee empowerment.
- Explain the rationale for empowering employees.
- Explain various inhibitors of empowerment and self-leadership.
- Explain the leader's role in empowerment.
- Explain how to achieve empowerment and self-leadership.
- Describe how to recognize empowered, self-leading employees.
- Demonstrate how to avoid empowerment traps.
- Summarize the lessons of selected leaders.

"Buffalo hunters used to slaughter the herd by finding and killing the leader. Once the leader was dead the rest of the herd stood around waiting for instructions that never came, and the hunters could (and did) exterminate them one by one."[1] Although your competitors are not likely to do something so drastic as kill your organization's head buffalo, there will be times when the leader is absent from the herd, so to speak. Whether the leader is away at a meeting, on vacation, or sick at home, employees should be able to continue their work at the same high level of performance as when he is present.

An organization cannot compete in a global environment if its employees—like the leaderless buffalo herd—simply stand around waiting for instructions when the leader is away. This is why the best leaders teach their employees self-leadership and then empower them to apply the concept in the leader's absence, as well as in his presence.

EMPLOYEE SELF-LEADERSHIP DEFINED

Self-leadership on the part of employees means taking the initiative in all situations to think and act in the best interests of the organization without waiting to be told to. The fact that employee self-leadership is practiced is a sign of a well-led organization. Employee self-leadership occurs when the following conditions exist:

- Leaders ensure that employees understand the big picture as well as their individual role and that of their team members in achieving it.
- Employees take responsibility for their individual performance as well as that of their team.
- Employees are empowered to apply the concept of self-leadership.

Ralph Stayer uses the analogy of a flock of geese flying in formation to illustrate his view of employee self-leadership: "I didn't want an organizational chart with traditional lines and boxes, but a 'V' of individuals who knew the common goal, took turns leading, and adjusted their structure to the task at hand. Geese fly in a wedge, for instance, but land in waves. Most important, each individual bird is responsible for its own performance."[2] When geese fly in a V formation, they take turns leading at the point of the V where the wind resistance is the most severe. After leading for a while, a goose will drop back to the end of the V, and another will take its place in the lead. Because each goose is expected to take responsibility for leading the V at some point in the migration, the analogy works well as an example of self-leadership.

EMPLOYEE EMPOWERMENT DEFINED

To understand self-leadership one must first understand empowerment. Defined simply, *empowerment* is "employee involvement that matters." Another good definition of empowerment is "the controlled transfer of authority to make decisions and take action." It's the difference between just having input and having input that is heard, seriously considered, and followed up on whether it is accepted or not.

Most employee involvement systems fail within the first year, regardless of whether they consist of suggestion systems, regularly scheduled brainstorming sessions, daily quality circles, one-on-one discussions between employees and a supervisor, or any combination of the various involvement methods. The reason is simple: they involve but do not empower the employees. Employees soon catch on to the difference between having input and having input that matters. Without empowerment, involvement is just another management tool that doesn't work.

EMPOWERMENT DOES NOT MEAN ABDICATION

It is not uncommon for traditional managers to view empowerment as an abdication of power. Such managers see empowerment as turning over control of the organization to the employees. In reality, this is hardly the case. Empowerment involves actively soliciting input from those closest to the work and giving careful thought to that input. It involves giving employees the right, within defined ranges, to take the initiative in solving problems and making improvements.

Pooling the collective mind of all people involved in a process, if done properly, will enhance rather than diminish a manager's power. It increases the likelihood that the information on which decision makers base their decisions is comprehensive and accurate. Managers do not abdicate their responsibility by adopting empowerment. Rather, they increase the likelihood of making the best possible decisions and thereby more effectively carry out their leadership responsibilities.

Leadership Tip

"Treat employees like partners and they will act like partners."[3]

—Fred Allen
Chairman, Pitney-Bowes Company

RATIONALE FOR EMPOWERING EMPLOYEES

Traditionally, working hard was seen as the surest way to succeed. With the advent of global competition and the simultaneous advent of automation, the key to success became not just working hard but also working smart. In many cases, decision makers in business and industry interpreted working smart as adopting high-tech systems and automated processes. These smarter technologies have made a difference in many cases. However, improved technology is just one aspect of working smarter, and it's a part that can be quickly neutralized when the competition adopts a similar or even better technology.

An aspect of working smart that is often missing in the modern workplace is involving and empowering employees in ways that take advantage of their creativity and promote independent thinking and initiative on their part. In other words, what's missing is empowerment. Creative thinking and initiative from as many employees as possible will increase the likelihood of better ideas, better decisions, better quality, better productivity, and, therefore, better competitiveness. The rationale for empowerment is that it represents the best way to bring the creativity and initiative of the best employees to bear on improving the organization's competitiveness.

Human beings are not robots or automatons. While working they observe, think, sense, and ponder. It is natural for a person to continually ask such questions as the following:

- Why is it done this way?
- How could it be done better?
- Will the customer want the product like this?

Asking such questions is an important step in making improvements. As employees ask questions, they also generate ideas for solutions, particularly when given the opportunity to regularly discuss their ideas in a setting that is positive, supportive, and mutually nurturing.

Empowerment is sometimes seen by experienced managers as just another name for participatory management. However, there is an important distinction between the two. Participatory management is about managers and supervisors asking for their employees' help. Empowerment is about getting employees to help themselves, each other, and the company. This is why empowerment can be so effective in helping maintain a high level of motivation among employees. It helps employees develop a sense of ownership of their jobs and of the company. This in turn leads to a greater willingness on the part of employees to make decisions, take risks in an effort to make improvements, and speak out when they disagree.

INHIBITORS OF EMPOWERMENT AND SELF-LEADERSHIP

The primary inhibitor of empowerment is resistance to change. Although most resistance comes from management, surprisingly there is sometimes resistance from employees and unions. Since there can be no self-leadership without empowerment, those who would lead should understand the most common inhibitors of empowerment and what to do about them.

Resistance from Employees and Unions

Some employees suffer from the "what's next" syndrome. In this syndrome, employees have experienced so many flash-in-the-pan management strategies that either did not work out or were not followed through on that they have become skeptical.

In addition to skepticism, there is the problem of inertia. Resistance to change is natural. Even positive change can be uncomfortable for employees because it involves new and unfamiliar territory. However, when recognized for what they are, skepticism and inertia can be overcome.

Unions can be another source of resistance when implementing empowerment. Because of the traditional adversarial relationship between organized labor and management, unions may be suspicious of management's motives in implementing empowerment. They might also resent an idea not originated by their own organization. However, union members' greatest concern is likely to be how empowerment will affect their future. If union leaders think it will diminish the need for their organization, they will throw up roadblocks.

A well-known case in business circles dealing with union resistance to empowerment and how to overcome it is the FPL Group (originally Florida Power & Light). This large utility company is well known for its excellent quality-improvement program. A major component in the quality program is employee empowerment. As good as the company's program is, union leadership at FPL Group initially resisted it. The reason for the union's resistance was failure by management to involve union leaders. What overcame the union's resistance was simple—FPL Group's management team stepped back, started over, and involved the union leadership in developing the program.

When union leaders saw that management intended to use empowerment and quality as strategies for improvement rather than as cleverly worded strategies for getting more out of employees without paying more for it, they became the program's most effective advocates. Whether unionized or not, employees should be involved from the outset in establishing and implementing the empowerment philosophy. When the concept is viewed as something employees and management do together, it will receive more support than when viewed as something management is "doing to us."

Resistance from Management

Even if employees and labor unions support empowerment, it will not work unless management makes a full and wholehearted commitment to it. Some companies attempt to implement empowerment without first making the necessary fundamental changes in organizational structure or management style.

The importance of management commitment cannot be overemphasized. Employees take their cues from management concerning what is important, to what the company is committed, how to behave, and all other aspects of the job.

Peter Grazier (*Before It's Too Late: Employee Involvement*, 1989) summarizes the reasons behind management resistance to empowerment as follows: insecurity, personal values, ego, management training, personality characteristics of managers, and exclusion of managers.[4]

Insecurity

An old adage states, Knowledge is power. By controlling access to knowledge as well as the day-to-day flow of knowledge, managers can maintain power over employees. Managers who view the workplace from an us-against-them perspective tend to be insecure about any initiative they perceive as diminishing their power.

Another source of management insecurity is accountability. Pooling the minds of employees to make workplace improvements is a sure way to identify problems, roadblocks, and inhibitors. Some managers fear they will be revealed as the culprit in such a process. The natural reaction of an individual who feels threatened is to resist the source of the threat. Managers are no different from anyone else in experiencing this feeling.

Personal Values

Many of today's managers have a dogmatic mind-set when it comes to working with employees. This means they think employees should do what they are told, when they are told, and how they are told. Such a value system does not promote empowerment. Managers who feel this way will resist empowerment as being inappropriate. They are likely to think, "There can be only one boss around here, and that boss is me."

Ego

People who become managers may be understandably proud of their status and protective of the perquisites that accompany it. Status appeals to the human ego, and ego-focused managers may project an "I am the boss" attitude. Such managers may have difficulty reining in their egos enough to be effective participants in an approach they view as encroachment on territory that should be exclusively theirs.

Management Training

Many of today's managers were educated and trained by modern disciples of Frederick Taylor, the father of scientific management. Taylor's followers, whether university professors or management trainers, tend to focus on applying scientific principles to the improvement of processes and technology. Less attention is given to people-oriented improvements. Writing about Taylor, Grazier says, "Management was continuously looking for methods of dealing with the 'labor problem' and Taylor's scientific approach permitted them to deal with work flow procedures and equipment improvements rather than the more complex issues of employee commitment and morale. Taylor believed that it was the 'experts' who solved problems in organizations. And the culmination of his work was a philosophy that clearly defined management's role as the 'thinker' and labor's role as the 'doer'; that is, management does the thinking, and labor does what management says."[5]

Attitudes like the one reflected in this quote are inappropriate in the modern workplace. Nevertheless, vestiges of Taylor's school of thought remain in organizations throughout the world. Actually, much of what Taylor professed about applying scientific principles in improving the workplace is valid and has gained new status with the advent of total quality. Function analysis, statistical process control (SPC), and just-in-time manufacturing (JIT)—all total quality concepts—are examples of science applied to workplace improvement. However, unlike Taylor's followers, proponents of total quality involve employees in the application of these scientific methods. Management schools that still cling, even subtly, to the management-as-thinkers and labor-as-doers philosophy produce graduates who might resist empowerment. Experienced managers who were schooled in this philosophy in years past and have practiced it throughout their careers also might resist empowerment.

Personality Characteristics of Managers

Old-school managers are often found to be more task oriented then people oriented. They tend to focus more on the task at hand and getting it done than on the people who actually perform the task.

Consider the example of Mary, a task-oriented manager. Before her promotion, she did most of the work associated with task accomplishment herself. Her dependence on and interaction with other employees was minimal. As a result, Mary's task-oriented personality served her and the company well. Now, as a manager, she is responsible for organizing work and getting it done by others.

In this new setting, she has found that the quality and quantity of employees' work can be affected by problems they are having. She has found that, in spite of her well-earned reputation for getting the job done, other employees have ideas of their own about workplace improvements, and they want their ideas to be heard and given serious consideration. Finally, Mary has learned that employees have feelings, egos, and personal agendas and that these things can affect their work. Managers such as Mary who have a strong task orientation are not likely to support efforts such as empowerment that call for a balanced attitude in which the manager is concerned with both tasks and people.

Exclusion of Supervisors and Mid-Managers

Empowerment is about involving personnel who will be affected by an idea or a decision in making the decision. This includes the first level of management (supervisors), mid-management, and executive management. Any manager or level of management excluded from the process can be expected to resist. Even with a full commitment from executives and enthusiastic support from employees, empowerment will not succeed if mid-managers and supervisors are excluded. Those who are excluded, even if they agree conceptually, may resist simply because they feel left out.

Workforce Readiness

An inhibitor of employee empowerment that receives little attention in the literature is workforce readiness. Empowerment will fail quickly if employees are not ready to be empowered. In fact, empowering employees who are not prepared for the responsibilities involved can be worse than not empowering them at all. On the other hand, lack of readiness—even though it may exist—should not be used as an excuse for failing to empower employees. The challenge to management is twofold: (1) determine whether the workforce is ready for empowerment; and (2) if it is not ready, get it ready.

How, then, does one know whether the workforce is ready for empowerment? One rule of thumb is that the more highly educated the workforce, the more ready its members will be for empowerment. Because well-educated people are accustomed to critical thinking, they are experienced in decision making, and they tend to make a point of being well informed concerning issues that affect their work. This does not mean, however, that less-educated employees should be excluded. Rather, it means that they may need to be prepared before being included.

In determining whether employees are ready for empowerment, ask the following questions:

■ Are the employees accustomed to critical thinking?

- Are the employees knowledgeable of the decision-making process and their role in it?
- Are employees fully informed of the "big picture" and where they fit into it?

Unless the answers to all three of these questions are yes, the workforce is not ready for empowerment. An empowered employee must be able to think critically. It should be second nature for an employee to ask such questions as the following: Is there a better way to do this? Why do we do it this way? Could the goal be accomplished some other way? Is there another way to look at this problem? Is this problem really an opportunity to improve things?

These are the types of questions that lead to continuous improvement of processes and effective solutions to problems. These are the sorts of questions that empowered employees should ask all the time about everything. Employees who are unaccustomed to asking questions such as these should be taught to do so before being empowered.

Employees should understand the decision-making process, both on a conceptual level and on a practical level (e.g., how decisions are made in the organization). Being empowered does not mean making decisions. Rather, it means being made a part of the decision-making process. Before empowering employees, you need to show them what empowerment will mean on a practical level. How will they be empowered? Where do they fit into the decision-making process? They also need to be aware of the boundaries. What decisions are they able to make themselves or within their work teams? Employees should know the answers to all of these questions before being empowered.

An employee who does not know where the organization is going will be unable to help it get there. Before empowering employees, you should educate them concerning the organization's strategic plan and their role relative to it. When employees can see the goal, they are better able to help the organization reach it.

Organizational Structure and Management Practices

Most resistance to empowerment is attitudinal, as the inhibitors explained so far show. However, a company's organizational structure and its management practices can also work against the successful implementation of empowerment. Before you attempt to implement empowerment, ask the following questions:

- How many layers of management are there between workers and decision makers?
- Does the employee performance-appraisal system encourage or discourage initiative and risk taking?

- Do management practices encourage employees to speak out against policies and procedures that inhibit quality and productivity?

People will become frustrated if their ideas have to work their way through a bureaucratic maze before reaching a decision maker. Prompt feedback on suggestions for improvement is essential to the success of empowerment. Too many layers of managers who can say no between employees and decision makers who can say yes will inhibit and eventually kill initiative on the part of employees.

Employees with initiative will occasionally make mistakes or try ideas that don't work. If this reflects negatively on their performance appraisals, initiative will be replaced by a play-it-safe approach. This also applies to constructive criticism of company policies and management practices. Are employees who offer constructive criticism considered problem solvers or troublemakers? Managers' attitudes toward constructive criticism will determine whether they receive any. A positive, open attitude in such cases is essential. The free flow of constructive criticism is a fundamental element of empowerment.

THE LEADER'S ROLE IN EMPOWERMENT

The leader's role in empowerment can be stated simply. It is to do everything necessary to ensure successful implementation and ongoing application of the concept. The three words that best describe the leader's role in empowerment are *commitment*, *leadership*, and *facilitation*. All the functions are required to break down the barriers and overcome the inherent resistance often associated with implementation of empowerment or with any other major change in the corporate culture.

Grazier describes the leader's role in empowerment as demonstrating the following support behaviors:[6]

- Exhibiting a supportive attitude
- Being a role model
- Being a trainer
- Being a facilitator
- Practicing management by walking around (MBWA)
- Taking quick action on recommendations
- Recognizing the accomplishments of employees

ACHIEVING EMPOWERMENT AND SELF-LEADERSHIP

Figure 9.1 shows the four broad steps in the implementation of empowerment. Creating a workplace environment that is positive toward and supportive of empowerment so that risk taking and individual initiative are

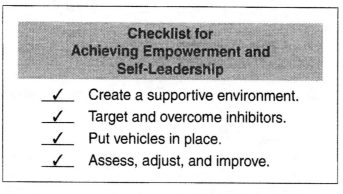

FIGURE 9.1 Achieving empowerment and employee self-leadership requires a systematic effort.

encouraged is critical. Targeting and overcoming inhibitors of empowerment is also critical. These two steps were discussed earlier in this chapter. The third and fourth steps are dealt with here.

A number of different types of vehicles can be used for soliciting employee input and channeling it to decision makers. Such vehicles range from simply walking around the workplace and asking employees for their input, to periodic brainstorming sessions, to regularly scheduled quality circles. Following are some widely used methods that are typically the most effective.

Brainstorming

With brainstorming, leaders serve as catalysts in drawing out group members. Participants are encouraged to share any idea that comes to mind. All ideas are considered valid. Participants are not allowed to make judgmental comments or to evaluate the suggestions made. Typically, one member of the group is asked to serve as a recorder. All ideas are recorded, preferably on a marker board, flip chart, or another medium that allows group members to review them continuously.

After all ideas have been recorded, the evaluation process begins. Participants are asked to go through the list one item at a time, weighing the relative merits of each. This process is repeated until the group narrows the choices to a specified number. For example, leaders may ask the group to reduce the number of alternatives to three, reserving the selection of the best of the three to themselves.

Brainstorming can be an effective vehicle for collecting employee input and feedback, particularly if leaders understand the weaknesses associated with it and how they can be overcome. Leaders interested in soliciting employee input through brainstorming should be familiar with the concepts of groupthink and groupshift. These two concepts can undermine the effectiveness of brainstorming and other group techniques.

Groupthink is the phenomenon that exists when people in a group focus more on reaching a decision than on making a good decision.[7] Several factors can contribute to groupthink, including the following: overly prescriptive group leadership, peer pressure for conformity, group isolation, and unskilled application of group decision-making techniques. Mel Schnake (*Human Relations*, 1990) recommends the following strategies for overcoming groupthink.[8]

- Encourage criticism.
- Encourage the development of several alternatives. Do not allow the group to rush to a hasty decision.
- Assign a member or members to play the role of devil's advocate.
- Include people who are not familiar with the issue.
- Hold last-chance meetings. When a decision is reached, arrange a follow-up meeting a few days later. After group members have had time to think things over, they may have second thoughts. Last-chance meetings give employees an opportunity to voice their second thoughts.

Groupshift is the phenomenon that exists when group members exaggerate their initial position, hoping that the eventual decision will be what they really want.[9] If group members get together prior to a meeting and decide to take an overly risky or unduly conservative view, this can be difficult to surmount. Leaders can help minimize the effects of groupshift by discouraging reinforcement of initial points of view and by assigning group members to serve as devil's advocate.

Nominal Group Technique

The nominal group technique (NGT) is a sophisticated form of brainstorming involving five steps (see Figure 9.2). In the first step, the leader states the problem and provides clarification if necessary to make sure all group members understand. In the second step, group members silently record their ideas. At this point, there is no discussion among group members. This strategy promotes free and open thinking unimpeded by judgmental comments, peer pressure, or negative nonverbal feedback.

In the third step, the ideas of individual members are made public by asking each member to share one idea with the group. The ideas are recorded on a marker board or flip chart. The process is repeated until all ideas have been recorded. Each idea is numbered. There is no discussion among group members during this step. Taking the ideas one at a time from group members ensures a mix of recorded ideas, making it more difficult for members to recall which idea belongs to which individual.

In the fourth step, recorded ideas are clarified to ensure that group members understand what is meant by each. A group member may be asked to

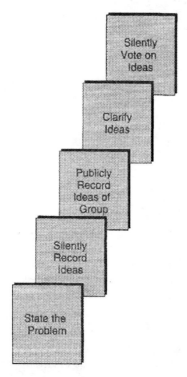

FIGURE 9.2   Nominal Group Technique (NGT).

explain an idea, but no comments or judgmental gestures are allowed from other members. The member clarifying the ideas is not allowed to make justifications. The goal in this step is simply to ensure that all ideas are clearly understood.

In the final step, the ideas are voted on silently. There are a number of ways to accomplish this. One simple technique is to ask all group members to record the numbers of their five favorite ideas on five separate 3 × 5 cards. Members then prioritize their five cards by assigning them a number ranging from 1 (worst idea) to 5 (best idea). The cards are collected and the points assigned to ideas are recorded on the marker board or flip chart. After this process has been accomplished for all five cards of all group members, the points are tallied. The idea receiving the most points is selected as the best idea.

Quality Circles

A quality circle is a group of employees that meets regularly to identify, recommend, and make workplace improvements. A key difference between quality circles and brainstorming is that quality circle members are volunteers who convene themselves and conduct their own meetings. Quality circles are an excellent example of employee self-leadership.

Brainstorming sessions are typically convened and conducted by a manager. A quality circle has a team leader who acts as a facilitator, and the group may use brainstorming, NGT, or other group techniques; however, the team leader is typically not a manager and may, in fact, be a different group member at each meeting. Quality circles meet regularly before, during, or after work to anticipate problems, propose workplace improvements, set goals, and make plans.

Walking and Talking

As mentioned earlier, simply walking around the workplace and talking with employees can be an effective way to solicit input. This approach is sometimes referred to as management by walking around (MBWA). An effective way to prompt employee input is to ask questions. This approach may be necessary to get the ball rolling, particularly when empowerment is still new and not yet fully accepted by employees. In such cases, it is important to ask the right questions. According to Joseph T. Straub, writing in *Supervisory Management*, managers should "ask open-ended, unbiased questions that respect . . . workers' views and draw them out."[10] Regardless of the vehicles used for soliciting employee input, leaders need to concentrate on practicing employee empowerment and promoting self-leadership.

Leadership Profile   Empowerment Through Training at Motorola [11]

One of the most effective empowerment tools available to technology companies is training. Training empowers employees to do what is expected of them in a changing environment. Said another way, an employee who does not know how to do what is expected is not empowered. At Motorola, ongoing training for employees at all levels is one of the company's highest priorities and the foundation of its empowerment effort. The company spends more than $50 million annually on employee training.

For example, professional development courses are provided for supervisors and middle managers on such topic as interpersonal skills, how to make effective presentations, holding effective meetings, and a variety of other subjects. This training ensures that all employees can complete their tasks as Motorola strives to stay ahead in the marketplace.

Assess, Adjust, Improve

Once an organization has created a supportive environment, targeted inhibitors and taken the necessary steps to overcome them, and put vehicles in place, the developmental work is done. There is, however, still much more work to be done—the work of maintaining. Supervisors and managers should continually assess the effectiveness of the environment. Is it as supportive as it should be? Is it becoming more supportive or less? Are there new inhibitors that need to be dealt with? Based on these on-going assessments, leaders adjust and make improvements. This process never ends.

HOW TO RECOGNIZE EMPOWERED, SELF-LEADING EMPLOYEES

There will always be managers and supervisors who resist the concept of empowerment out of fear of losing control or losing their authority.[12] Some will give lip service to empowerment while continuing to do things the same way they always have. How can an organization's leaders know that their empowerment efforts are working? In other words, how does one recognize an empowered employee who practices self-leadership? The following comparisons will help leaders determine whether they have empowered employees or employees who are required to do things as they have always been done, by supervisors who talk about empowerment but do not really believe in it.

Taking initiative.   Empowered employees who are self-leaders will take the initiative in ambiguous situations and define problems in ways that enable further analysis and lead to positive action. Employees who are not empowered and are not self-leaders will wait for someone in authority to define the problem and initiate action.

Identifying opportunities.   Empowered, self-leading employees will identify opportunities for improvement in the problems that arise. Employees who are not empowered and are not self-leaders may solve the immediate problem but fail to identify ways to improve processes and prevent future occurrences of the problems.

Thinking critically.   Empowered, self-leading employees feel free to think critically, question the status quo, and challenge assumptions. Employees who are not empowered and are not self-leaders are more likely to take information at face value without testing its validity.

Building consensus.   Empowered, self-leading employees build consensus among all stakeholders within groups and across functional areas. Employees who are not empowered and are not self-leaders are more likely to simply look to a higher authority to mandate a decision.

No matter what an organization's leaders say about empowerment and employee self-leadership, and no matter how elaborate the systems put in place to promote these concepts, employees are not empowered and are not self-leaders until they are willing and able to take the initiative when action is needed, identify opportunities for continual improvement in the problems that occur, build consensus for a given action or decision, and think critically when considering actions, decisions, and assumptions.

Leadership Tip

"All great leaders have possessed the capacity of believing in the capabilities and talents of others. Those who are always disdainful of subordinates, who constantly denigrate their work, who always compare their efforts unfavorably with their own will wind up leading no one but themselves."[13]

AVOIDING EMPOWERMENT TRAPS

According to Kyle Dover ("Avoiding Empowerment Traps"[14]), despite their potential, "empowerment programs often fall victim to the very structural and cultural problems that made them desirable in the first place. Many managers view empowerment as a threat and continue to measure their value by the authority they wield. Meanwhile, some employees mistake empowerment for discretionary authority—the power to decide things unilaterally—and lack the collaborative skills that management neglects, or refuses, to teach them. Others resist the need to assume more power and cling to a comfortable dependence on authority."[15] Organizations should avoid falling into the following empowerment traps:

Defining power as discretion and self-reliance.   Empowerment and self-leadership give employees the authority to think critically, make decisions within controlled parameters, and participate fully. They do not give them the authority to act unilaterally on the basis of their own discretion. Empowerment is about being a full participant in a team process. It is not about being a self-reliant loner.

Failing to properly define empowerment for mid-managers and supervisors.   When mid-managers and supervisors view empowerment as a loss of authority, status, or power, they tend to resist it. They may feel as if they worked long and hard to gain a position of authority, only to be asked to hand that authority over to employees. Ensuring that mid-managers and supervisors understand the issue of controlled delegation of authority is critical if empowerment is to achieve its potential as a performance-improving concept.

Assuming employees have the skills to be empowered.   You should not ask an employee who does not know how to think critically, make decisions, identify improvement opportunities, or build consensus to do so. Before implementing an empowerment program, it is important to assess the skill level of employees and provide them with the training and mentoring needed to make effective use of their status as empowered employees.

Getting impatient.   Making the transition from the traditional approach to empowerment and employee self-leadership represents a major change, a change that can take time. Processes have to be changed, employees have to be trained, fears have to be worked through, and sufficient time must pass to allow people to grow accustomed to a new culture. Expecting immediate results is unrealistic. Leaders of empowerment efforts should be prepared to deal with and overcome organizational impatience.

LESSONS ON EMPOWERMENT FROM SELECTED LEADERS

Following are excerpts from the lives of several leaders that exemplify some of the empowerment principles set forth in this chapter. The leaders selected for inclusion here are Ulysses S. Grant, Peter McKinnon, and Fred Smith.

Ulysses S. Grant

As a general, Grant wanted soldiers at all levels who would think—and he empowered them to do so. Commenting on those he led during the Civil War, Grant once said, "Our armies were composed of men who knew what they were fighting for ... and so necessarily must have been more than equal to men who fought merely because they were brave and because they were thoroughly drilled and inured to hardship."[16] What Grant said in this case was that his soldiers saw the big picture, knew where they fit into it, and, as a result, could use their minds to determine the best course of action in any situation they faced. According to Al Kaltman, the empowerment lesson taught by the career of Ulysses S. Grant is as follows:

> Empowerment begins with knowledge, which goes beyond job training.
> No matter how well you teach your people to do their jobs, if they don't
> understand the organization's mission and the important role they play in
> carrying it out, all you will have is people who act like non-thinking
> robots, and you will always be outperformed by any competitors who
> empower their staff to think for themselves.[17]

Grant's actions taught this lesson many times in his career. During his Civil War campaign against the river fortress of Vicksburg, Mississippi, Grant had been stymied for months. The Confederate batteries that lined the Mississippi River had successfully repulsed every strategic and tactical move the Union army had attempted. Finally, Grant's army was able to get past the guns of Vicksburg and attack the city from behind. But the heavily fortified city was too well defended, and Grant's initial assaults were repulsed. His best option was to lay siege to Vicksburg, but Grant had a problem. He had an insufficient number of large cannon and mortars—weapons essential to a successful siege. Grant challenged his officers to find a solution, and he empowered them to think, innovate, and improvise. This they did, and before long the problem had been solved. "Wooden ones (cannons) were made by taking logs of the toughest wood that could be found, boring them out for six or twelve pound shells and binding them with strong iron bands. . . . Shells were successfully thrown from them into the trenches of the enemy."[18]

Peter McKinnon[19]

Although banks certainly use some of the most modern technologies, they are not technology companies—at least not in the context of this book. However, Peter McKinnon of the National Bank of Australia has a lesson to teach those who would lead that is so applicable that he has been included here. McKinnon was given responsibility for establishing a leadership development program that would "focus principally on three dimensions of leadership: contact, clarity, and impact. Contact refers to competencies that involve a leader's ability to be in touch with themselves, their businesses, and their team. Clarity has to do with the idea that leaders must be pathfinders who set new directions for their organizations and teams; they need to provide clarity about future goals and directions for their organizations. Impact refers to whether the actions and ideas of the leaders influence others."[20]

The impact component of the program is designed to develop leaders who will be (1) flexible and adaptable, (2) committed to making a difference, (3) able to communicate with impact, and (4) able to adopt a clear service orientation. The clarity component is designed to develop leaders who will be (1) intellectually robust, (2) continually involved in learning more about business, and (3) effective change agents. The contact component is designed to develop leaders who will (1) have a strong self-regard, (2) act with integrity, and (3) bring out the best in people. The leadership lesson taught by Peter McKinnon is as follows:

> Leaders can and must learn to develop the ability and willingness to, among other things, empower their followers to think and act in the best interest of the organization.

This lesson may not be readily apparent to the reader upon first consideration of McKinnon's leadership development program—but look closer. Several of the individual elements of the program have an empowerment aspect. Leaders who "communicate with impact" are those who show their direct reports the big picture—an essential first step in empowerment. Leaders cannot be "effective change agents" unless they empower their followers to think, innovate, and act in ways that promote positive change. Finally, leaders cannot "bring out the best in people" unless they are willing to empower them.

Fred Smith

Fred Smith is well known in the world of business as the person who conceptualized one of the most successful companies in the world in a paper he was assigned to write in college.[21] The paper received only a mediocre grade, but under Fred Smith's dynamic leadership the company conceptualized in it—Federal Express—has consistently made an A. One of the ways Smith ensures that Federal Express employees are fully empowered to think, act, and innovate for continual improvement is to periodically survey them about their managers. The empowerment lesson taught by Fred Smith is as follows:

> If a CEO wants managers and supervisors to use employee empowerment as a leadership strategy, he must expect, monitor, measure, and reinforce the desired behavior.

To show managers and supervisors at Federal Express that he expects employees to be empowered for self-leadership, Smith monitors, measures, and reinforces the concept using the FedEx Survey. The survey contains questions such as the following:

- My manager makes sure I know what is expected of me.
- My manager takes the time to listen to my concerns.
- I have the freedom I need to do my job well.
- The concerns identified in the last survey have been satisfactorily addressed.

Employees are asked to respond to questions such as these as well as others using a 6-point scale. The possible responses are as follows: Strongly Agree, Agree, Sometimes Agree/Disagree, Disagree, Strongly Disagree, and Undecided/Don't Know. Clearly, at Federal Express those who would lead are expected to empower.

Summary

1. Self-leadership means taking the initiative to think and act in the best interests of the organization. Empowerment means engaging employees in the thinking processes of an organization in ways that matter. Empowerment means having input that is heard and seriously considered. Empowerment requires a change in organizational culture, but it does not mean that managers abdicate their responsibility or authority.

2. The rationale for empowerment is that it is the best way to increase creative thinking and initiative on the part of employees. This, in turn, is an excellent way to enhance an organization's competitiveness. Another aspect of the rationale for empowerment is that it can be an outstanding motivator.

3. The primary inhibitor of empowerment is resistance to change. Resistance might come from employees, unions, or management. Management-related inhibitors include insecurity, personal values, ego, management training, personality characteristics, exclusion, organizational structure, and management practices.

4. The leader's role in empowerment is best described as commitment, leadership, and facilitation. The kinds of support managers can provide include having a supportive attitude, role modeling, training, facilitating, employing MBWA, taking quick action on recommendations, and recognizing the accomplishments of employees.

5. Achieving empowerment and self-leadership requires four broad steps: creating a supportive environment; targeting and overcoming inhibitors; putting the vehicles in place; and assessing, adjusting, and improving. Vehicles include brainstorming, nominal group technique, quality circles, and walking and talking.

6. The empowerment lesson from Ulysses S. Grant is that employees must understand the big picture in order to truly help the organization. The lesson of Peter McKinnon is that leaders must develop the ability and willingness to empower their followers. The lesson of Fred Smith is that the CEO must expect, monitor, measure, and reinforce the empowerment behavior of managers.

7. A workforce that is ready for empowerment is accustomed to critical thinking, understands the decision-making process, and knows where it fits into the big picture.

Key Terms and Concepts

Assess

Brainstorming

Coaching

Commitment

Empowerment

Facilitation

Fear of rejection

Groupshift

Groupthink

Hidden barriers

Initiative

Negative behavior

Nominal group technique

Review Questions

1. Define the terms *empowerment* and *self-leadership*.
2. Explain the following statement: "Achievement of empowerment and employee self-leadership requires change in the corporate culture."
3. Give a brief rationale for empowerment.
4. What is the relationship between empowerment and motivation?
5. List three inhibitors of empowerment and employees' self-leadership and how they can be overcome.
6. Explain the various root causes of management resistance to empowerment.
7. In what ways can an organization's structure and management practices inhibit empowerment?
8. Describe the leader's role in empowerment.
9. Describe how to use brainstorming to promote empowerment.
10. What is a quality circle?
11. Describe the concept of MBWA.
12. Explain the concept of workforce readiness as it relates to empowerment.
13. Summarize the empowerment lessons from Ulysses S. Grant, Peter McKinnon, and Fred Smith.

LEADERSHIP SIMULATION CASES

The following simulations are provided to generate additional thought and discussion about the principles of leadership explained in this chapter. Readers are encouraged to consider how the situations presented in these cases might apply to them and to discuss the cases with other leaders and leadership candidates.

CASE 9.1 Resistance to Empowerment

Bruce Harolle, CEO of Container Tech, Inc., did not mince words in responding to a suggestion from his chief engineer that he consider empowering employees at CTI to think more. "You can forget all this nonsense about employee empowerment. In this company, I think and employees work. That's why we are the leader in our field." "Bruce, I know we are the market leader right now, and we have been since you took over as CEO," said the exasperated chief engineer. "But that could change anytime. The last three jobs we bid had competition from Indonesia, Argentina, and Korea. I don't remember that ever happening before." He knew that convincing Harolle that employee empowerment and self-leadership were important would be a tough sell. But he also knew that with global competition becoming the norm, CTI had to change—and that meant Harolle had to change.

Discussion Questions

1. Have you ever worked in a situation in which your supervisor had a don't-think-just-do-as-you-are-told attitude? If so, what effect did this attitude have on your performance? On the performance of the organization?
2. If you were CTI's chief engineer, how would you go about convincing Harolle that he and the company needed to adopt the employee empowerment philosophy?

CASE 9.2 Forget Your Ideas—Just Do It My Way

Amanda Parker had worked her way up from a maintenance position to the head of the printed circuit board fabrication department. Parker has an exceptional ability to focus on a task, break it down into its component parts, arrange the parts in a logical sequence, and tackle each part of the task in the proper order until the entire task is completed. Before becoming a supervisor, Parker had been the most productive printed circuit board fabrication technician the company had ever employed. Now she managed her old shop.

As a manager, things are not working out for Parker. She has found working with people much more challenging than working with tasks and processes. Often subordinates think they have a better idea for how to perform certain tasks. Parker would rather they keep their ideas to themselves and do things the way she tells them to. After all, wasn't she the company's best printed circuit board fabricator before moving up into management?

Discussion Questions

1. Why might it be difficult for a manager who used to be a technician to let employees do their job?
2. What personal inhibitors will Parker have to overcome in order to be willing and able to empower her employees for self-leadership?

CASE 9.3 New Management, No Commitment

Employees of Brown & Sons, Inc., were accustomed to giving their input before decisions were made and to being asked for their opinions whenever a problem arose. In fact, one of the reasons Brown & Sons has always been such a competitive company is that employee empowerment and self-leadership are fully integrated into the corporate culture. This point was made clear by Randall Brown, the sole proprietor and CEO of Brown & Sons, when at the age of 79 he decided to sell the company and retire. The new owners claimed to understand and agree with Brown's philosophy of employee empowerment—that is for about three months.

Now, six months after purchasing Brown & Sons, the new owners have become almost dictatorial in their attitudes toward employee input and feedback. Predictably, the company's performance has fallen off sharply. To make matters worse, as the company's performance continues to decline, the new management team becomes more and more prescriptive. If things don't change soon, Brown and Sons might be facing Chapter 11 proceedings.

Discussion Questions

1. Have you ever been in a situation in which a change of management was followed by a change of philosophy? Was the change for the better or worse? Explain.

2. If you worked at Brown & Sons, what would you say to try to convince your new executive manager of the efficacy of employee empowerment and self-leadership?

Endnotes

[1] Ralph Stayer, "How I Learned to Let My Workers Lead," HBR OnPoint, Harvard Business School Publishing Corporation, Product Number 8172, 2001, 33.

[2] Ralph Stayer, 33.

[3] Louis E. Boone, Quotable Business, 2nd ed. (New York: Random House, 1999), 95.

[4] Peter Grazier, Before It's Too Late: Employee Involvement (Chadds Ford, PA: Teambuilding, 1989), 8.

[5] Peter Grazier, 103–104.

[6] Peter Grazier, 129–142.

[7] D. G. Myers and H. Lamm, "The Group Polarization Phenomenon," Psychological Bulletin 80 (1971): 251–270.

[8] Mel E. Schnake, Human Relations (New York: Macmillan, 1990), 285–286.

[9] R.D. Clark, "Group-Induced Shift Toward Risk: A Critical Appraisal," Psychological Bulletin 80 (1971): 251–270.

[10] J. T. Straub, "Ask Questions First to Solve the Right Problems," *Supervisory Management* (October 1991): 7.

[11] Belohlav, J. A. *Championship Mangement: An Action Model for High Performance* (Cambridge, MA: Productivity Press, 1990), 122.

[12] Kyle Dover, "Avoiding Empowerment Traps," *Management Review* (January 1999): 53.

[13] *Bits and Pieces* (July 25, 1992): 19.

[14] Kyle Dover, 51–55.

[15] Kyle Dover, 51–52.

[16] Al Kaltman, *Cigars, Whiskey and Winning—Leadership Lessons from General Ulysses S. Grant* (Upper Saddle River, NJ: Prentice Hall, 1998), 13.

[17] Al Kaltman, 13.

[18] Al Kaltman, 134.

[19] Jay A. Conger and Beth Benjamin, *Building Leaders* (San Francisco: Jossey-Bass, 1999), 73–78.

[20] Jay A. Conger and Beth Benjamin, 73.

[21] Jay A. Conger and Beth Benjamin, 100.

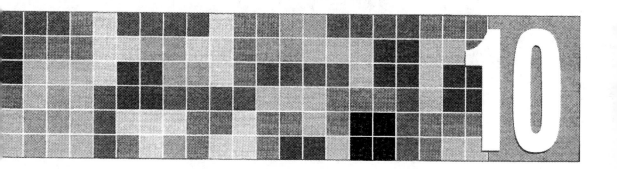

Be an Effective Conflict Manager and Consensus Builder

"It is much easier to be critical than to be correct."[1]

—Benjamin Disraeli

OBJECTIVES

- Define internal politics.
- Explain the concepts of power and politics at work.
- Describe the methods of internal politicians.
- Explain the impact of internal politics on competitiveness.
- Describe how to control internal politics in organizations.
- Explain how to overcome negativity in organizations.
- Explain how to overcome territorial behavior in organizations.
- Explain how to manage conflict in an organization.

MANAGING CONFLICT IN ORGANIZATIONS

Conflict is a normal and unavoidable aspect of the highly competitive modern workplace. One of the human relations skills needed by people in such a setting is the ability to disagree with colleagues without being disagreeable. However, even if most members of an organization have this skill, there is no guarantee that conflicts will not arise among workers. When people work together, no matter how committed they are to a common goal, human conflict is going to occur. Consequently, leaders must be proficient in resolving conflict.

Causes of Workplace Conflict

The most common causes of workplace conflict are predictable. They include those shown in Figure 10.1. .

Limited resources often lead to conflict in the workplace. It is not uncommon for an organization to have fewer resources (funds, supplies, personnel, time, equipment, etc.) than might be needed to complete a job. When this happens, who gets the resources and in what amounts? *Incompatible goals* often lead to conflict, and incompatibility of goals is inherent in the workplace. For example, conflicts between engineering and manufacturing are common. The goal of engineering is to design a product that meets the customers' needs. The goal of manufacturing is to produce a high-quality product as inexpensively as possible. In an attempt to satisfy the customer, engineering might create a design that is difficult to manufacture economically. The result? Conflict.

Role ambiguity can also lead to conflict by blurring "turf lines." This makes it difficult to know who is responsible and who has authority. *Different values* can lead to conflict. For example, if one group values job security and another values maximum profits, the potential for conflict exists.

FIGURE 10.1   Common causes of workplace conflict.

> ### Causes of Workplace Conflict
>
> - Limited resources
> - Incompatible goals
> - Role ambiguity
> - Different values
> - Different perspectives
> - Communication problems

Different perceptions can lead to conflict. How people perceive a given situation depends on their background, values, beliefs, and individual circumstances. Because these factors are sure to differ among both individuals and groups, particularly in an increasingly diverse workplace, perception problems are not uncommon.

The final predictable cause of conflict is *communication*. Effective communication is difficult at best. Improving the communication skills of employees at all levels should be an ongoing goal of leaders. Because communication will never be perfect, communication-based conflict should be expected.

How People React to Conflict

To deal with conflict effectively, leaders need to understand how people react to conflict. According to management expert K. W. Thomas, the ways in which people react to conflict can be summarized as competing, accommodating, compromising, collaborating, or avoiding.[2]

A typical reaction to conflict is *competition*, in which one party attempts to win while making the other lose. The opposite reaction to conflict is *accommodation*. In this reaction, one person puts the needs of the other first. *Compromise* is a reaction in which the two opposing sides attempt to work out a solution that helps both to the extent possible. *Collaboration* involves both sides working together to find an acceptable solution for both. *Avoidance* involves shrinking away from conflict. This reaction is seen in people who are not comfortable facing conflict and dealing with it.

In some situations, a particular reaction to conflict is more appropriate than another. People who are responsible for resolving conflict need to understand what is and what is not an appropriate reaction to conflict. Thomas ("Toward Multi-Dimensional Values in Technology"[3]) has summarized the various situations in which specific reactions to conflict are appropriate.

- Competing is appropriate when quick action is vital or when important but potentially unpopular actions must be taken.
- Collaborating is appropriate when the objective is to learn or to work through feelings that are interfering with interpersonal relationships.
- Avoiding is appropriate when you perceive no chance of satisfying your concerns or when you desire to let people cool down and have time to regain a positive perspective.
- Accommodating is appropriate when you are outmatched and losing anyway or when harmony and stability are more important than the issue at hand.

How Conflict Should Be Handled

Leaders have two responsibilities regarding conflicts: (1) conflict resolution and (2) conflict stimulation. When conflict is present, leaders need to resolve it in ways that serve the organization's long-term best interests. This will keep conflict from becoming a detriment to performance. Where no conflict exists, leaders may need to stimulate it to keep the organization from becoming stale and stagnant.[4] Both of these concepts taken together are known as conflict management.

D. Tjosvold sets forth the following guidelines that can be used by leaders in attempting to resolve conflict:[5]

- Determine how important the issue is to all people involved.
- Determine whether all people involved are willing and able to discuss the issue in a positive manner.
- Select a private place where the issue can be discussed confidentially by everyone involved.
- Make sure that both sides understand that they are responsible for both the problem and the solution.
- Solicit opening comments from both sides. Let them express their concerns, feelings, ideas, and thoughts, but in a nonaccusatory manner.
- Guide participants toward a clear and specific definition of the problem.
- Encourage participants to propose solutions. Examine the problem from a variety of different perspectives and discuss any and all solutions proposed.
- Evaluate the costs versus the gains (cost-benefit analysis) of all proposed solutions and discuss them openly. Choose the best solution.
- Reflect on the issue and discuss the conflict-resolution process. Encourage participants to express their opinions as to how the process might be improved.

How and When Conflict Should Be Stimulated

Occasionally an organization will have too little conflict. Such organizations tend to be those in which employees have become too comfortable and management has effectively suppressed freethinking, innovation, and creativity. When this occurs, stagnation generally results. Stagnant organizations need to be shaken up before they die. Leaders can do this by stimulating positive conflict or conflict that is aimed at revitalizing the organization. S. P. Robbins says that a yes response to any of the following questions suggests a need for conflict stimulation:[6]

- Are you surrounded by employees who always agree with you and tell you only what you want to hear?

- Are your employees afraid to admit they need help or that they've made mistakes?

- Do decision makers focus more on reaching agreement than on arriving at the best decision?

- Do managers focus more on getting along with others than on accomplishing objectives?

- Do managers place more emphasis on not hurting feelings than on making quality decisions?

- Do managers place more emphasis on being popular than on high job performance and competitiveness?

- Are employees highly resistant to change?

- Is the turnover rate unusually low?

- Do employees, supervisors, and managers avoid proposing new ideas?

Each time one of these questions is answered in the affirmative, it is an indication that conflict may need to be stimulated. It may be possible to have a vital, energetic, developing, improving organization without conflict, but this isn't likely to happen. Innovation, creativity, and the change inherent in continual improvement typically breed conflict. Therefore, the absence of conflict can also be an indication of the absence of vitality. Therefore, managers need to know how to stimulate positive conflict.

According to Robbins, the techniques for stimulating conflict fall into three categories: improving communication, altering organizational structure, and changing behavior.[7]

- Improving communication will ensure a free flow of ideas at all levels. Open communication will introduce a daily agitation factor that will ensure against stagnation while at the same time providing a mechanism for effectively dealing with the resultant conflict.

- Altering organizational structure in ways that involve employees in making decisions that affect them and that empower them will help prevent stagnation. Employees in organizations that are structured to give them a voice will use that voice. The result will be positive conflict.

- Changing behavior may be necessary, particularly in organizations that have traditionally suppressed and discouraged conflict rather than dealing with it. Leaders who find themselves in such situations may find the following procedure helpful: (1) identify the types of behaviors you want employees to exhibit, (2) communicate with employees so that they understand what is expected, (3) reinforce the desired behavior, and (4) handle conflict as it emerges using the procedures set forth in the previous section.

Communication in Conflict Situations

The point has been made that human conflict in the workplace is normal, to be expected, and, in certain instances, to be promoted. In managing conflict—which in essence means resolving conflict when it is having negative effects and promoting conflict when doing so might help avoid stagnation—communication is critical.

The following guidelines concerning the use of communication in managing conflict can be helpful:[8]

The *initial attitude of those involved in the conflict can predetermine the outcome.*   This means that a person who enters into a situation spoiling for a fight will probably get one. Communication prior to such a situation aimed at convincing either or both parties to view it as an opportunity to cooperatively solve a problem can help predetermine a positive outcome.

When possible, conflict guidelines should be in place before conflicts occur.   It is not uncommon for conflict to be exacerbated by disagreements over how it should be resolved. Before entering into a situation in which conflict might occur, make sure all parties understand how decisions will be made, who has the right to give input, and what issues are irrelevant.

Assessing blame should not be allowed.   It is predictable that two people in a conflict situation will blame each other. If human interaction is allowed to get hung up on the rocks and shoals of blame, it will never move forward. The approach that says, "We have a problem. How can we work together to solve it?" is more likely to result in a positive solution than arguing over who is to blame.

"More of the same" solutions should be eliminated.   When a particular strategy for resolving conflict is tried but proves to be ineffective, don't continue using it. Some managers get stuck on a particular approach and stay with it even when the approach clearly doesn't work. Try something new instead of using "more of the same" solutions.

Maintain trust by keeping promises.   Trust is fundamental to effective leadership. It is especially important in managing conflict. Trust is difficult to win but easy to lose. Conflict cannot be effectively managed by someone who is not trusted. Consequently, leaders must keep their promises, and in so doing build trust among employees.

Each of these strategies depends on communication. Communication that is open, frank, tactful, continual, and inclusive can do more than anything else to ensure that conflict is properly managed in the workplace.

INTERNAL POLITICS DEFINED

A historic example of internal politics at its worst occurred during World War II. Pearl Harbor and the Philippines had fallen before the Japanese onslaught in the Pacific, and it looked as if New Guinea and Australia would be next. The Allied forces were in a real bind. On the one hand, they couldn't allow these two countries to fall into enemy hands. But on the other hand, they couldn't pull enough troops away from the European and Atlantic theaters of operation to stop Japan.

In desperation, President Franklin D. Roosevelt ordered General Douglas MacArthur to leave Bataan, the tiny sliver of land in the Philippines to which MacArthur and his American and Filipino troops had clung tenaciously for several months. MacArthur and his beleaguered troops were holding out against great odds to give the United States time to recover from the devastating blow suffered at Pearl Harbor. This was a monumental decision, because if Bataan fell—which it surely would without MacArthur's presence—the Japanese would add the Philippines to their list of conquests. Reluctantly, MacArthur and a small staff slipped away under cover of darkness aboard patrol torpedo (PT) boats of the American navy. Slipping through the Japanese blockade, they eventually reached Australia.

In short order, MacArthur rallied the dangerously understaffed and underequipped military units available to him. Relying on innovative strategy, raw courage, and stubborn determination, MacArthur and his troops made their stand and stopped the Japanese juggernaut on the Owen Stanley mountain range in New Guinea. From that point on, MacArthur conducted one of the most brilliant military campaigns in the annals of war, eventually pushing a stronger, healthier, better-fed, better-equipped Japanese military out of the southwest Pacific.

The Japanese army and navy were well trained, well led, and fanatically determined, but they weren't MacArthur's most difficult foe. Also working against him was the insidious scourge of internal politics. To conduct the type of island-hopping warfare that was necessary in the southwest Pacific, MacArthur needed the U.S. Army, Navy, Marines, and Air Corps, as well as military units from both Australia and New Zealand, to work together in a closely coordinated, collaborative effort. As it turns out, trying to achieve coordination and collaboration among the disparate groups under his command was General MacArthur's greatest challenge and most persistent frustration.

Rivalries among different branches of the U.S. military, as well as within individual branches of the military, were legendary at the time Japan attacked Pearl Harbor. The personal ambitions of military leaders, service loyalties, disagreements over the allocation of resources, and jealousy relating to who was in command of what were all ongoing sources of problems that created almost as much trouble for MacArthur as did the Japanese. In the

European theater, General Eisenhower faced the same frustrations daily as he tried to coordinate the military forces of the various Allied nations.

In a meeting with Army Chief of Staff George C. Marshall, MacArthur vented his frustration. Recalling this meeting in his memoirs, MacArthur said, "I felt it fantastic, to say the least, that interservice rivalry or personal ambitions were allowed to interfere with winning the war."[9] This same kind of situation can be found in almost any organization. Internal politics is a natural, if unfortunate, outgrowth of human nature.

Politics, in general, is the art and science of wielding influence in such a way as to gain advantage. One usually thinks of politics in relation to influencing government. However, politics as a means of influencing outcomes is not limited to government enterprises. In fact, one will find politics practiced with great skill in virtually every type of organization. Internal politics, as practiced in organizations, can be defined as follows:

> Internal politics consists of the games people play to promote decisions that are based on criteria other than merit. Internal politics manifests itself in a number of different behaviors, all of which amount to individuals or groups within an organization putting their interest ahead of those of the overall organization.

Internal politics is not necessarily an inherently bad concept. For example, internal politics practiced for the purpose of furthering the interests of the overall organization would be acceptable. The problem with the concept is that it is rarely practiced in a positive way. Even though internal politicians invariably claim they are acting in the best interests of the organization, most typically have their own interests at heart in every action they take.

In a worst-case scenario, internal politicians are people who put self-interest ahead of organizational interests. In a better-case scenario, internal politicians are people who put not self-interest but the interests of their divisions, departments, or teams ahead of those of the overall organization. It is this—the self-serving nature of the concept as it is typically practiced—that makes internal politics such a negative phenomenon in organizations. It undermines collaboration, trust, and unity of purpose—all fundamental elements of global competitiveness.

Leadership Tip

Individuals in organizations—frequently the organization's key decision makers—can become so intent on fulfilling their ambitions, satisfying their personal needs, and feeding their individual egos that they lose sight of what's best for the organization that employs them.

—The Author

POWER AND POLITICS

Power is the ability to exert influence. Power is neither inherently positive nor inherently negative. It is a concept that can cut either way, depending on how it is used. Power, properly applied in an organization, is used to move the organization closer to the realization of its vision. Power, improperly applied in an organization, is used to advance an agenda other than that of the overall organization. This is the way in which internal politicians use power. The five different sources of power in an organization are personal, position, capability, reward, and coercive power. Internal politicians may use any or all of these sources of power to advance their personal agendas.

■ *Personal power.* Personal power is the power of an individual's personality. Individuals with personal power are generally persuasive or charismatic. They tend to have strong beliefs, an aura of confidence, and an air of determination. In the military such people are said to have the intangible attribute of "command presence." Their personalities appeal to certain people—their followers—on an emotional level.

■ *Position power.* Position power is that bestowed officially by higher authority. People with the authority to hire and fire, allocate resources, evaluate performance, and make decisions that affect the jobs of others have position power. People with position power may or may not be adept at using their power to influence others. Weak leaders in positions of authority often find that their position alone is not enough to ensure their influence in an organization. An important aspect of power is knowing how to use it and being willing to do so. For example, a supervisor who is unwilling to confront nonproductive employees will not be able to influence their behavior for the better, in spite of having position power.

■ *Capability power.* Capability power comes from having special knowledge, skills, or talents that are important to an organization and in short supply. In any organization there are critical tasks that are fundamental to the organization's success. People who can perform these tasks often gain power. In some instances their power will extend well beyond that which might be expected of a person in their position.

■ *Reward power.* Reward power comes from the authority to control, administer, or withhold something of value to others. Often the something in question is money. People in a position to give or withhold money from others in organizations have reward power. In addition to money and resources, rewards that might be granted or withheld include recognition, promotion, positive feedback, and inclusion in a group. The reward of inclusion is one of the favorite tools of the internal politician. People have a natural desire to be part of a group. Consequently, the ability to exclude others gives an individual power over those who want to be associated with a given group.

■ *Coercive power.* People who have the ability to punish others or subject them to unpleasant circumstances have coercive power. Coercive power is founded in fear, and its application is based on threats. The threat—whether implicit or explicit—is to punish those who don't respond as expected. In an organizational setting, the coercive threat usually has to do with the victim's job security, work schedule, or pay. The threat can also be one of physical abuse, although this is less common in the workplace than are threats to job security. The threat of ostracism is also a commonly used coercive tactic. It involves threatening to remove or exclude an individual from a group with which he or she wants to be associated.

INTERNAL POLITICIANS AND THEIR METHODS

Internal politicians have many of the characteristics of special interest groups. Special interest groups consist of people who share narrowly focused common goals. Internal politicians are individuals with narrowly focused interests—namely, their own. Special interest groups seek to gain advantage so as to influence governmental decision making. Internal politicians seek to gain advantage so as to influence organizational decision making. With such strong similarities, it should come as no surprise that internal politicians and special interest groups use many of the same methods. The most widely used of these methods are

- Lobbying
- Building coalitions
- Applying harassment and pressure
- Electioneering
- Gossiping and spreading rumors

Lobbying

According to Alan Gitelson, Robert Dudley, and Melvin Dubnick, "Lobbying is the act of trying to influence government decision makers. Named after the public rooms in which it first took place, lobbying now goes on in hearing rooms, offices, and restaurants—any spot where a lobbyist can gain a hearing and effectively present a case."[10] Legend has it that the term *lobbying* is based on the fact that many of the earliest attempts to influence members of Congress occurred in the lobby of the Willard Hotel and other hotels in Washington, D.C.

Governmental lobbyists use favors, financial contributions, and information to influence government officials. Internal politicians, when lobbying, use similar tactics. By doing favors for people in positions to help them, internal politicians hope to curry favor. Ideally, they will establish a

---

Checklist of
Lobbying Tactics

- Contacting people formally (by appointment) to present a personal point of view.

- Engaging people in informal discussions (over lunch, on the golf course, in the hall, etc.) and presenting a personal point of view

- Providing carefully screened information on a selective basis.

- Doing favors to establish quid pro quo relationships.

- Helping lighten the workload of selected people.

- Applying pressure directly to individuals.

- Applying pressure through third parties.

- Exploiting personal relationships.

FIGURE 10.2   Lobbying tactics of internal politicians.

quid pro quo relationship with someone in a position of influence. Although they don't make financial contributions, internal politicians do have their version of this concept. They might contribute to easing the workload of, or solving a problem for, someone with whom they hope to gain favor.

Sharing information is another widely practiced lobbying tactic. There is an old saying that knowledge is power. By providing information to carefully selected people, internal politicians attempt to endear themselves. Figure 10.2 lists lobbying tactics commonly used by internal politicians.

Doing favors, making contributions, and providing information are not inherently negative activities—quite the contrary. What transforms these otherwise positive activities into negative endeavors is their misuse. If these things are done with the best interests of the organization in mind, they are admirable activities. However, when done for the purpose of advancing a personal agenda at the expense of or without sufficient consideration for the organization's needs, they become negative.

The information-sharing aspect of lobbying is frequently the most misused of the various lobbying tactics. Information provided for lobbying purposes is carefully shaded in favor of the information provider. An internal politician is not going to volunteer information that doesn't serve his or her purpose. This does not mean that internal politicians necessarily lie, nor that they even need to do so. Rather, it means that they carefully control the information they provide, and to whom it is provided, so as to gain the greatest possible benefit.

Building Coalitions

A coalition is a group of diverse people brought together by a common interest. In governmental politics, coalitions are formed to elect individuals to office, keep other individuals from being elected, secure budget appropriations, and pass legislation. Organizational coalitions are formed for various reasons, such as getting selected individuals promoted, ensuring that others are not promoted, securing resources, guaranteeing the adoption of favorable policies or procedures, and fostering a favorable organizational structure. The individuals or individual groups that make up a coalition may have nothing in common except the simple cause that brought them together. This fact gave rise to the old adage *Politics makes strange bedfellows*. Consequently, once a cause has been satisfied, the coalition typically dissolves. In its place others will form as interests, conditions, and circumstances change.

For the sake of illustration, suppose that the marketing and accounting departments of a hypothetical organization have never gotten along. Marketing personnel in this organization think that their colleagues from accounting are shortsighted, tightfisted, and don't understand that the organization must spend money to make money. Accounting personnel think that their colleagues from marketing are a bunch of high-rolling big spenders who refuse to work within a budget no matter how large it is. Then one day the organization's management team announces its plan to purchase a building across the street from its existing facility. By retaining its existing facility and relocating some of its personnel to the new building across the street, the company will gain badly needed work space. The only downside is that at least one department is going to have to move to the new building.

After analyzing space requirements, the executive team decides that either the engineering department alone or marketing and accounting together must move. It is not a good time for engineering to be disrupted by relocating, because the company has just received a large contract that is engineering intensive and has a "short fuse." There is no problem with marketing and accounting moving except that they are comfortable where they are and don't want to move. The current building that houses their offices has some amenities the new building won't have (covered parking, a cafeteria). Occupants of the new building will have to park their cars in a lot that is exposed to the weather and walk across the street to the old building to use the cafeteria.

Sensing that they are about to lose some of their valued perquisites, the accounting and marketing vice presidents, along with their respective staffs, form a coalition to lobby against moving. While engineering personnel are busy working on the organization's new contract, the accounting and marketing departments mount an effective lobbying campaign to have engineering relocated to the new building. Their lobbying efforts pay off, and the engineering department is moved across the street. Unfortunately, the disruption causes the department to fall behind in its work, and the organization's new contract goes over schedule. Late fees are assessed, and the relationship with a valuable new customer gets off to a bad start.

Applying Pressure and Harassment

In governmental politics, when pressure is applied, there is an implicit threat from voters: "Do what I ask or I won't vote for you." From lobbyists the threat is more along the lines of "Vote as we ask or lose our financial support." From colleagues in Congress, the unspoken message is "Support my bill or else I won't support yours." In organizations pressure is applied differently, but the implicit threat is still there. Examples of messages and tactics used by internal politicians in organizations to apply pressure are as follows:

- Help me out or you will be socially ostracized by your peers.
- Help me out and you will be part of the crowd.
- Help me out or something you don't want known will be revealed.
- If you help me out, I'll help you when I win. If you don't, you'll be left out when I win.
- Help me out or something undesirable will happen to someone you care about.
- Help me out and something good will happen to someone you care about.

The following scenario illustrates how internal politicians use pressure to serve their self-interests. John Brown is the purchasing agent for Orlando A&M University. Because he has a master's degree in accounting, Brown is able to earn extra income teaching night classes in freshman accounting. With twins on the way, Brown and his wife need the extra income.

In addition to his job as a purchasing agent, Brown chairs the university's staff development committee, the committee that allocates the funds used by faculty and staff members to attend professional conferences and to participate in professional development activities. There are always more requests for funds than there are funds available. Consequently, Brown and his committee have established some ironclad rules about the number of activities that will be funded for a given individual within a specified time frame. These rules ensure that the largest possible number of university employees get an opportunity to participate in professional development activities.

As chair of the development committee, Brown is accustomed to being in the "hot seat" when someone wants an activity funded and the committee cannot comply. But Brown has never been pressured as hard as he is currently being pressured by Amos Andrews, chair of the Department of Business and Accounting. The committee, sticking to the rules, has turned down a request from Andrews. After Andrews exhausted the list of ploys typically used by people trying to influence the committee, he began to get desperate.

That's when Brown really began to feel the pressure. Working through other faculty members in the business and accounting department, Andrews made sure Brown knew that the extra income he earned by teaching night classes in accounting was in jeopardy. The message given to Brown was clear: If Andrews's request for a waiver of the committee's rules is not approved, Brown has taught his last freshman accounting course. Brown needs the extra income from teaching, but he cannot approve Andrews's request without bumping Maxine Denny from a conference that is very important to the social science department. Brown is in a bind, and he is feeling the pressure.

This is just one example of the many ways that pressure can be applied in the workplace by internal politicians. There are many other ways that internal politicians can and do use personal pressure to advance their individual agendas.

Electioneering

In governmental politics, electioneering means participating in the election process. Participation can take many different forms, including raising money for candidates, making contributions, and getting out the vote. Of course, the purpose of electioneering is to ensure that a certain candidate is elected. Electioneering in an organization is a similar process.

Internal politicians use electioneering tactics to ensure that selected individuals are promoted, that the *right* people are appointed to prestigious committees, and that selected people are chosen to chair important committees and task forces. The following example illustrates electioneering as it might be used in an organizational setting.

Tim Jones is in a bind. In just six months he will be promoted to a planning position at the corporate office, unless, that is, someone at corporate learns the truth about his division. Brierfield Products, Inc. (BPI), is undergoing a partial restructuring. As a division director, Jones is a key player. However, he hasn't played the leadership role the company needs him and its other directors to play.

It's not that Jones is opposed to the restructuring; it's just that he had hoped to be off to his new position before having to bother with it. "After all," he thought, "why go through all the trouble when I won't be around to enjoy the benefits?" This type of thinking has led Jones to procrastinate. As a result, his division is lagging behind the company's other divisions. Jones has belatedly decided to get started, but he has a long way to go to catch up, and the monitoring visit is scheduled to occur in just one week.

When he made the decision to skate through his last six months, leaving the work of the restructuring to his successor, Jones had not known that corporate would conduct monitoring visits. Now a monitoring visit was right around the corner. If he didn't do something soon, not only would his promotion be lost but he'd also be lucky to keep his current job. But Jones isn't finished yet. He hadn't gotten to be one of the company's youngest division directors by accident. He possesses considerable skills as an internal politician, and it will be his political skills that will save him.

It was while scanning the list of personnel appointed by corporate to serve as monitors that Jones saw his chance for salvation. The third person on the list was Jake Burns. Jones couldn't believe his luck. He and Burns went way back, and better yet, Burns owed him. Now all he had to do was make sure that Jake Burns was selected as the monitor for his division. Jones began electioneering in earnest, pulling out all the stops. He made telephone calls, got other people to make telephone calls, applied pressure, made promises, made threats, and called in favors. By the time Jones was done, Jake Burns had been chosen as the monitor for the upcoming visit. Jones could finally relax. Jake Burns would write an appropriately worded, appropriately positive—albeit misleading—monitoring report.

Gossiping and Spreading Rumors

One of the most pernicious weapons in the arsenal of the internal politician is doubt. Doubt can be created effectively by using gossip to spread rumors about a targeted individual or group. When used by internal politicians, rumors and gossip are not the harmless chitchat variety. Rather, they are intentional, coldly calculated attempts to advance the agenda of one individual or group at the expense of another.

Rumors and gossip have the greatest impact when they cast doubt on an individual relative to high-priority organizational values. The following scenario demonstrates how rumormongering and gossip can be used by internal politicians to gain an advantage.

Patricia Chitwood is both ambitious and smart. She knows that her company, Drake Services, Inc., places a high priority on ethics, and she plans to use this fact to her advantage. Chitwood and her colleague Pamela McGraw have both applied for the soon-to-be-vacant position of regional sales manager at Drake Services. Like Chitwood, McGraw is good. In terms of both credentials and performance, the two sales reps could be twins. As things stand now, the race for the promotion is dead even. But Chitwood has a plan. At lunch today she will start a rumor that is sure to sow seeds of doubt about McGraw's ethics. Nothing major—just a few whispered comments and well-placed winks concerning McGraw's expense account. Within an hour, the office grapevine will be buzzing. Within a day, the CEO of Drake Services will be wondering about the ethics of one of his best sales reps. Within a week, Chitwood should be the new regional sales manager.

There is no question that Patricia Chitwood is a resourceful internal politician. However, her shortsighted methods may cost her in the long run. The rumor Chitwood started about her rival is likely to make Pamela McGraw's position with Drake Services tenuous at best. Even if McGraw isn't fired, she will probably leave; and with her record, landing a position with a competitor wouldn't be difficult. As regional sales manager, the last thing Chitwood will need is to have her best sales rep, Pamela McGraw, joining forces with the competition.

Leadership Profile　　Preventing Conflict and Internal Politics at Federal Express

One of the most effective ways to prevent conflict and internal politics on the job is to develop a written statement that explains how personnel are expected to treat each other.[11] The following expectations come from the corporate philosophy of Federal Express concerning human relations at work:

- Giving employee considerations a high priority when developing corporate programs and policies, when acquiring and designing facilities, equipment, and systems, and when scheduling and arranging work.

- Involving each employee as a valuable team member in all corporate activities. After all, who knows more about how the job should be done than those doing it?

- Being dedicated to promotions from within unless the needed skills cannot be found in existing employee ranks.

- Spending the time and effort necessary to manage the personnel issues—especially training and coaching.

- Maintaining outstanding communications and making available any information requested that is not personal, privileged, or controlled by government regulation.

- Enacting progressive programs such as our Open Door, Survey-Feedback-Action, and Guaranteed Fair Treatment Procedure/EEO Complaint Process policies to ensure that problems are solved and no individual is subjected to unreasonable or capricious treatment.

- Treating every single employee with respect and dignity.[12]

IMPACT OF INTERNAL POLITICS ON COMPETITIVENESS

The approach that is the opposite of internal politics is collaboration. The rationale for collaboration can be found in the negative impact internal politics can have on an organization. How successful would the U.S. military have been in the war with Iraq if our soldiers had shot at each other instead of the enemy? This farfetched scenario is analogous to what happens in organizations suffering from the scourge of internal politics.

Individuals in organizations—frequently the organizations' key decision makers—sometimes become so intent on fulfilling their personal ambitions, satisfying their individual needs, or feeding their own egos that they lose sight of what is best for the organization that employs them.

Internal politics can affect an organization in the same way that cancer affects an individual. Both start covertly inside the victim, often remaining invisible until serious damage is done, and both can spread quickly. Organizations stricken with the disease of internal politics ultimately suffer the effects shown in Figure 10.3.

An organization's morale suffers when infighting, buck-passing, and rumormongering—all of which invariably result from the practice of internal politics—are allowed to become part of the dominant corporate culture. Decisions that are questionable at best, and even potentially unsound, are not uncommon in organizations that condone internal politics. Any time decisions are based on criteria other than what is best for the organization, the organization suffers.

Internal politics invariably lead to counterproductive internal competition. Ideally, the only competition in which an organization would engage is market competition. Internal cooperation in the pursuit of a common purpose serves an organization better in the long run than internal competition among its own departments and employees. Organizations can devote their time and energy to the battle of the marketplace, or they can devote it to internal battles, but not to both. Resources are finite; time and energy wasted on internal squabbling are resources that could have been used to improve performance.

Organizations suffering from the ill effects of internal politics often lose the best and brightest employees. If a work team is low-performing and does not want to do better, its members may drive out anyone who tries to improve. Their methods range from peer pressure to harassment to outright ostracism. As they become increasingly frustrated by decisions based on politics rather than merit, employees with the most marketable credentials often show their dissatisfaction by leaving. Those who stay tend to fall into one of two categories: (1) employees who stay because weak credentials make it difficult for them to find a better job and (2) employees who give in to reality and become internal politicians themselves.

> ## Checklist of the
> ## Effects of Internal Politics
>
> - *Loss of morale* due to infighting, buck passing, and rumormongering.
> - *Questionable decisions* made for reasons other than what is best for the organization
> - *Counterproductive internal competition* that saps the organization of its competitive energy
> - *Loss of the best and brightest employees* as they make a statement about their dissatisfaction by leaving
> - *Perpetuation of outdated processes, procedures, and technologies* as internal politics is used to promote organizational inertia by those opposed to change
> - *Constant conflict* as the political machinations of one group are countered by those of others
> - *Loss of quality, competitiveness, and customers* as the organization's focus is diverted from what really matters

FIGURE 10.3   Internal politics can undermine an organization's ability to compete.

Internal politics tend to perpetuate outdated processes, procedures, and technologies because the tactics of the internal politician are ideally suited for opposing change. Change comes hard for most people. Psychological comfort with the status quo is inherent in the human condition. When internal politics become part of the corporate culture, organizations find it even more difficult than usual to make the changes necessary to stay competitive. With a little lobbying, some electioneering, and just the right amount of pressure wisely applied, the natural resistance of people to change can be magnified exponentially by internal politicians opposed to change. When this happens, the employees of an organization gain the psychological comfort associated with the status quo, but the organization loses the competitive edge associated with change.

Internal politics invariably multiply both the frequency and intensity of conflict in an organization. Infighting, backbiting, and ill will are antithetical to performance and competitiveness. Competitiveness requires unity of purpose and a trusting, mutually supportive work environment. Such environments cannot be maintained in the face of constant conflict that occurs on a personal rather than a professional level. All of the individual deleterious effects of internal politics, when taken together, have the cumulative effect of diverting an organization's attention from what really matters. This can occur to such an extent that the organization's quality suffers. When this happens, the organization loses its ability to satisfy and retain customers, and in turn its ability to succeed in a competitive marketplace.

CONTROLLING INTERNAL POLITICS IN ORGANIZATIONS

How can a leader convince people in an organization—all of whom have their own interests, ambitions, and egos—to put aside the natural inclination to practice internal politics and, instead, practice collaboration? Interestingly, trying to control internal politics in organizations is a lot like trying to prevent overeating by individuals. Both involve finding ways to subdue human nature, both require persistent effort, and both demand constant vigilance. Controlling internal politics in an organization requires a comprehensive effort involving all employees. Such an effort should have at minimum the components shown in Figure 10.4.

Strategic Planning Component

Controlling internal politics begins with the organization's strategic plan. One of the keys to controlling internal politics over the long term is creating a cultural expectation that all decisions will be based on what is best from the perspective of the organization's strategic plan. If employees are to make all decisions based on this criterion, they have to know the organization's vision, mission, core values, and broad objectives. In other words, if their behavior and decisions are supposed to support the strategic plan, employees have to know the plan. The strategic planning components of an

---

Components of an Internal Politics Prevention Program

- Strategic planning component
- Leadership component
- Reward and recognition component
- Performance-appraisal component
- Customer focus component
- Conflict management component
- Cultural component

FIGURE 10.4   Main components of an internal politics prevention program.

organization's effort to control internal politics should have at least the following elements:

- A procedure for explaining the strategic plan to all employees and how it is to be used in guiding all decisions and actions in the organization.
- A method for building a core value into the strategic plan that conveys the message that collaboration is the expected approach in the organization.

Explaining the Strategic Plan and Using It in Decision Making

All employees should have a copy of the strategic plan, the plan should be thoroughly explained, and employees should be given ample opportunity to ask questions and seek clarification about the plan. In other words, it's not enough for employees to have the plan—they need to understand it.

In explaining the strategic plan to employees, leaders should make sure that they convey the following message:

> Everything we do in this organization is to be guided by one criterion: support of the strategic plan. A good decision is one that supports accomplishing what is set forth in the strategic plan. A decision that does not meet this criterion is a bad decision. Good policy, good procedures, and good work practices are those that support the strategic plan. Others are unacceptable. Consequently, there is no room in the organization for advancing personal agendas or promoting self-interest to the detriment of organizational interests. In short, internal politics has no place in this organization.

Conveying this message to all employees, in conjunction with explaining the strategic plan, sets the proper tone, establishes the proper expectations, and brings the issue of internal politics out into the open. Setting the proper tone and establishing expectations are essential to developing an organizational culture that does not promote or condone internal politics. Bringing the issue out into the open removes the shroud of secrecy on which internal politicians thrive.

Collaboration as a Guiding Principle

One of the most important components of a strategic plan is the one that contains the organization's core values. The organization's core values are its guiding principles. These principles explain in writing what is most important to the organization and how it intends to do business. One of the core values that guides an organization should be collaboration. An example of such a core value is as follows:

ABC Company places a high priority on collaboration among all employees at all levels. We base all policies, procedures, practices, and decisions on what is best for the organization, rather than what serves the personal interests, agendas, or ambitions of individuals or individual units within the company.

If employees at all levels know that collaboration is a high priority, it becomes more difficult for them to play the games collectively known as internal politics. This principle can go a long way toward thwarting the practice of internal politics in the organization.

Leadership Component

A fundamental premise of leadership is setting a positive example. Leaders must be consistent role models of the behavior they expect of employees. If a manager practices internal politics, employees will respond in kind. Consequently, it is important that managers be seen using the organization's strategic plan as the basis for all actions and that they insist all employees follow suit.

Setting an example goes beyond just adopting policies and making decisions based on what best supports the organization's strategic plan. It also involves refusing to condone—explicitly or implicitly—counterproductive behavior by employees (e.g., rumormongering or gossiping). In fact, counterproductive behavior such as rumormongering and gossiping give leaders excellent opportunities for demonstrating the point that internal politics is not condoned. By openly and consistently refusing to gossip, spread rumors, or respond to either, managers can help take away two of the most potent weapons of internal politicians.

Reward and Recognition Component

If you want to promote a certain type of approach—for example, collaboration—reward it and recognize employees who practice it. This is a simple but effective leadership principle. Unfortunately, it's a principle that is preached more than it is practiced. One of the most frequent systemic mistakes made in organizations is failing to match up management expectations with reward and recognition systems. Perhaps the most common example of this failing can be found in organizations that expect teamwork but still maintain a reward system that is based on individual performance. Managers should examine their organization's incentives carefully to identify ways in which internal politics is rewarded, either directly or indirectly. The most obvious question is "What happens to employees at any level who are found to practice internal politics?" Another question is "Does the organization provide incentives that promote employee collaboration, and, if so, what are those incentives?" A well-designed reward and recognition system will simultaneously provide disincentives to internal politics and incentives that promote collaboration. Disincentives that can work against internal politics include negative performance appraisals, verbal warnings, and written reprimands. Incentives can be both formal and informal, and there are hundreds of both varieties.

Performance-Appraisal Component

The periodic performance appraisal is how most organizations formally let employees know how they are doing. Consequently, one or more of the criteria in an organization's performance-appraisal instrument should relate to collaboration. Here are two examples of these types of criteria.

1. This employee bases all actions on what is best for the organization.
   - ❏ Always
   - ❏ Sometimes
   - ❏ Usually
   - ❏ Never
2. What is this employee's collaboration rating?

   - ❏ Excellent
   - ❏ Above average
   - ❏ Average
   - ❏ Poor

Making collaboration an issue in performance appraisals ties it directly to pay and promotions. This is critical. Remember that internal politics is driven by self-interest.

Tying pay and promotion to an employee's willingness to practice collaboration means that internal politics no longer serves his or her self-interest.

Leadership Tip

"Some of the most effective sources of recognition cost nothing at all. A sincere word of thanks from the right person at the right time can mean more to an employee than a raise, a formal award, or a whole wall of certificates and plaques."[13]

—Bob Nelson

Customer Focus Component

Customer focus is a fundamental cornerstone of competitiveness. In organizations with a customer focus, quality is defined by customers, and the organization's strategic plan is written from the perspective of attracting, satisfying, and retaining customers. A customer focus is achieved by partnering with customers.

When an organization partners with its customers, it brings them into the decision-making process by actively seeking their input and feedback. Input, remember, consists of customer recommendations given *before* the decision is made. Feedback is customer information given *after* the decision is made. Input influences the decision that is made. Feedback evaluates the quality of the decision that was made. In organizations that factor customer input and feedback into the decision-making process, it is difficult for internal politicians to play their games. When decisions are driven by customer preference, they cannot be driven by politics.

The full benefit of a customer focus—from the perspective of preventing internal politics—is gained by ensuring that all employees are thoroughly informed concerning customer input and feedback. In this way customer needs and preferences become the critical criteria by which the viability of policies, procedures, practices, and decisions can be judged. Such criteria make it difficult for internal politicians to play their games. Even the most accomplished internal politician will find it difficult to justify recommending decisions that run counter to customer preferences.

Conflict Management Component

Internal politics tends to generate counterproductive conflict. This is one of the reasons that leaders in organizations should do what is necessary to prevent internal politics. However, it is important to distinguish between conflict and counterproductive conflict. Not all conflict is bad. In fact, properly managed conflict that has the improvement of products, processes, people, or the work environment as its source is positive conflict.

Counterproductive conflict—the type associated with internal politics—occurs when people in organizations behave in ways that work against the interests of the overall organization. This type of conflict is often characterized by deceitfulness, vindictiveness, and personal rancor. Productive conflict occurs when right-minded, well-meaning people disagree, without being disagreeable, concerning the best way to support the organization's strategic plan.

Positive conflict leads to discussion, debate, and give-and-take interaction among people whose only goal is to find the best solution or make the best decision. This type of interaction exposes the viewpoints of all participants to careful scrutiny and judges the merits of all arguments by applying criteria that are accepted by all stakeholders. By putting every point of view under the microscope of group scrutiny, weaknesses in arguments can be identified, the issue can be viewed from all possible angles, and, ultimately, the best solution can be identified. Contrast this approach with that of internal politicians whose only goal is to promote self-interest. Internal politicians review only that information that serves their interests. Internal politicians reveal only that information that serves their interests, while concealing any information that might weaken their case. They lobby, practice electioneering, and apply pressure to influence people to make decisions based on criteria other than merit.

By practicing conflict management, managers in an organization can make it difficult for internal politicians to play their games. Conflict management has the following components:

- Establishing conflict guidelines
- Helping all employees develop conflict prevention and resolution skills
- Helping all employees develop anger management skills
- Stimulating and facilitating productive conflict

Establishing Conflict Guidelines

Conflict guidelines set forth the ground rules for discussing and debating different viewpoints, ideas, and opinions concerning how best to accomplish the organization's vision, mission, and broad objectives. Figure 10.5 is an example of an organization's conflict guidelines. Guidelines such as these should be developed with a broad base of employee involvement from all levels in the organization.

Developing Conflict Prevention and Resolution Skills

If leaders are going to expect employees to disagree without being disagreeable, they are going to have to ensure that all employees are skilled in the art and science of conflict resolution. The second guideline in Figure 10.5 is an acknowledgment of human nature. It takes advanced human relations skills and constant effort to disagree without being disagreeable. Few people are born with this ability, but fortunately it can be learned. The following strategies are based on a three-phase model developed by Tom Rusk and described in his book *The Power of Ethical Persuasion*.[14]

Explore the other person's viewpoint.   A good start in preventing conflict can be made by acknowledging the importance of the other person's point of view. Begin the discussion by giving the other party an opportunity to present her point of view, and, listening carefully, say, "Your viewpoint is important to me, and I'm going to hear you out." The following strategies will help make this phase of the discussion more positive and productive:

Conflict Guidelines

Av-Tech, Inc., encourages discussion and debate among employees at all levels concerning better ways to continually improve the quality of our products, processes, people, and work environment. This type of intellectual interaction, if properly handled, will result in better ideas, policies, procedures, practices, and decisions. However, human nature is such that conflict can easily get out of hand, take on personal connotations, and become counterproductive. Consequently, in order to promote productive conflict, Av-Tech has adopted the following guidelines. These guidelines are to be followed by all employees at all levels.

- The criterion to be applied when discussing or debating any point of contention is as follows: Which recommendation is most likely to move our company closer to realizing the strategic vision?

- Disagree, but don't be disagreeable. If the debate becomes too hot, stop and give all parties an opportunity to cool off before continuing. Apply your conflict resolution skills and anger management skills. Remember, even when we disagree about how to get there, we are all trying to reach the same destination.

- Justify your point of view by tying it to either the strategic plan or customer input and feedback and require others to follow suit.

- In any discussion of differing points of view, ask yourself the following question: Am I just trying to win the debate for the sake of winning (ego), or is my point of view really the most valid?

FIGURE 10.5   Conflict guidelines for Av-Tech, Inc.

1. Establish that your goal at this point in the discussion is mutual understanding.
2. Elicit the other person's complete point of view.
3. Listen nonjudgmentally and do not interrupt.
4. Ask for clarification if necessary.

5. Paraphrase the other person's point of view and restate it to show that you understand.

6. Ask the other person to correct your understanding if it appears to be off base or incomplete.

Explain your viewpoint.   After you accurately and fully understand the other person's point of view, present your own. The following strategies will help make this phase of the discussion more positive and productive:

1. Ask for the same type of fair hearing for your point of view that you gave the other party.

2. Describe how the other's person's point of view affects you. Don't point the finger of blame or be defensive. Explain your reactions objectively, keeping the discussion on a professional level.

3. Explain your point of view accurately and completely.

4. Ask the other party to paraphrase and restate what you have said.

5. Correct the other party's understanding, if necessary.

6. Review and compare the two positions (yours and that of the other party). Describe the fundamental differences between the two points of view and ask the other party to do the same.

Agree on a resolution.   Once both viewpoints have been explained and are understood, it is time to move to the resolution phase. This is the phase in which both parties attempt to come to an agreement. It is also the phase in which both parties may discover that they cannot agree. Agreeing to disagree—in an agreeable manner—is an acceptable outcome. The following strategies will help make this phase of the discussion more positive and productive:

1. Reaffirm the mutual understanding of the situation.

2. Confirm that both parties are ready and willing to consider options for coming to an acceptable resolution.

3. If it appears that differences cannot be resolved to the satisfaction of both parties, try one or more of the following strategies:

   - Take time out to reflect and then try again.
   - Agree to third-party arbitration or neutral mediation.
   - Agree to a compromise solution.
   - Take turns suggesting alternative solutions.
   - Yield once your position has been thoroughly stated and is understood. The eventual result may vindicate your position.
   - Agree to disagree while still respecting each other's position.

Developing Anger Management Skills

It is difficult, if not impossible, to keep conflict positive when anger enters the picture. If individuals in an organization are going to be encouraged to question, discuss, debate, and even disagree, they must know how to manage their anger. Anger is an intense emotional reaction to conflict in which self-control may be lost. Anger occurs when people feel that one or more of their fundamental needs are being threatened. These needs include the following:

- Need for approval
- Need to be valued
- Need to be appreciated
- Need to be in control
- Need for self-esteem

When one or more of these needs is threatened, a normal human response is to become angry. An angry person can respond in one of four ways:

1. *Attacking.* With this response the source of the threat is attacked, usually verbally. For example, when someone disagrees with you (threatens your need for approval), you might attack by questioning his veracity or credentials.

2. *Retaliating.* With this response you fight fire with fire, so to speak. Whatever is given, you give back. For example, if someone calls your suggestion ridiculous (threatens your need to be valued), you might retaliate by calling his suggestion dumb.

3. *Isolating.* This response is the opposite of venting. With the isolation response, you internalize your anger, find a place where you can be alone, and simmer. The childhood version of this response was to go to your room and pout. For example, when someone fails to even acknowledge your suggestion (threatens your need to be appreciated), you might swallow your anger, return to your office, and boil over in private.

4. *Coping.* This is the most positive response to anger. Coping does not mean that you don't become angry. Rather, it means that even when you do, you control your emotions instead of letting them control you. A person who copes well with anger is a person who, in spite of his anger, stays in control. All employees at all levels of an organization need to be able to cope with their anger. The following strategies will help employees manage their anger by becoming better at coping.

- Avoid the use of anger-inducing words and phrases, including the following: *but, you should, you made me, always, never, I can't, you can't.*

- Admit that others don't make you angry but that you allow yourself to become angry. You are responsible for your emotions and your responses to them.
- Don't let pride get in the way of progress. You don't have to be right every time.
- Drop your defenses when dealing with people. Be open and honest.
- Relate to other people as equals. Regardless of position or rank, you are no better than them and they are no better than you.
- Avoid the human tendency to rationalize your angry responses. You are responsible and accountable for your behavior.

Stimulating and Facilitating Productive Conflict

Sycophantic agreement with the boss has no place in a competitive organization. Ideas, suggestions, and proposals should be subjected to careful, even intense scrutiny. Consequently, productive conflict is not only allowed but also promoted. Productive conflict consists of genuine, harmonious disagreement concerning the best way to solve a problem.

Productive conflict is productive because the only agenda being advanced is the good of the organization. With productive conflict no hidden agendas or political machinations are at work. All parties are attempting to reach the same destination; the disagreement has to do with how best to get there. Because there are no hidden agendas, all parties are open to questions, challenges, and constructive criticism. In addition, all parties agree on the criteria by which their ideas will be judged.

In a competitive organization, leaders actually stimulate discussion and debate (productive conflict) if they think a proposal is moving down the track too fast unimpeded by careful scrutiny. Productive conflict is stimulated using methods such as the following:

- Openly communicating the message "We want ideas and constructive criticism of ideas. We believe discussion and debate sharpen our ideas."
- Playing "devil's advocate" and teaching employees to play this role.
- Requiring employees with suggestions to identify the downside when making them.

Cultural Component

There are many definitions for the term *culture*. The *American Heritage College Dictionary* defines it as follows: "The totality of socially transmitted behavior patterns, arts, beliefs, institutions, and all other products of human thought."[15] The key part of this definition is "socially transmitted behavior patterns." As applied to an organization, the concept means the way things are done in the organization. In other words, an organization's culture is the everyday manifestation of its actual beliefs. The concept grows out of *actual* beliefs as opposed to *written* beliefs. An organization's culture *should* be the everyday manifestation of the core values found in its strategic plan.

However, some organizations are guilty of practicing a set of beliefs that differ from those written down as core values. Culture cannot be mandated. Rather, it develops over time based on actions, not words. This is why it is so important to live out the organization's professed beliefs daily.

If collaboration is a high priority, it should be promoted, modeled, rewarded, and reinforced daily at all levels in the organization. Correspondingly, internal politics must be seen to be ineffective and detrimental. For every incentive to collaborate, there should also be a disincentive to play political games. The ultimate disincentive is social pressure, which is why establishing a collaborative culture is so important and beneficial. It ensures that social pressure, the most effective enforcer of culture, works *for* collaboration instead of against it.

By applying the various strategies described in this section every day, leaders can make collaboration a fundamental part of their organization's culture. When this happens, social pressure within the organization will keep the practice of internal politics under control.

NEGATIVITY IN ORGANIZATIONS

Negativity is any behavior by any employee at any level that works against the optimum performance of the organization. The motivation behind negativity can be as varied as the employees who manifest it. However, most negative behavior can be categorized. Common categories of negative behavior are as follows:

- Control disputes
- Territorial disputes (boundaries)
- Dependence and independence issues
- Need for attention and responsibility
- Authority
- Loyalty issues

Recognizing Negativity in the Organization

Leaders should be constantly alert to signs of negativity in the workplace because negativity is contagious. It can spread throughout an organization quickly, dampening morale and inhibiting performance. Following are symptoms of the negativity syndrome that leaders should watch for.

■ *"I can't" attitudes.* People in an organization that is committed to continuous improvement have "can do" attitudes. If "I can't" is being heard regularly, negativity has crept into the organization.

■ *"They" mentality.* In high-performance organizations, employees say "we" when talking about their employer. If employees refer to the organization as "they," negativity has gained a foothold.

■ *Critical conversation.* In high-performance organizations, coffee-break conversation is about positive work-related topics or topics of personal interest. When conversation is typically critical, negative, and judgmental, negativity has set in. Some managers subscribe to the philosophy that employees are not happy unless they are complaining. This is a dangerous attitude. Positive, improvement-oriented employees will complain to their supervisor about conditions that inhibit performance, but they don't sit around criticizing and whining during coffee breaks.

■ *Blame fixing.* In a high-performance organization, employees fix problems, not blame. If blame fixing and finger-pointing are common in an organization, negativity is at work.

Overcoming Negativity

Leaders who identify negativity in their organizations should take the appropriate steps to eliminate it. Here are some strategies that can be used to overcome negativity in organizations:

■ *Communicate.* Frequent, ongoing, effective communication is the best defense against negativity in organizations, and it is the best tool for overcoming negativity that has already set in. Organizational communication can be made more effective using the following strategies: acknowledge innovation, suggestions, and concerns; share information so that all employees are informed; encourage open, frank discussion during meetings; celebrate milestones; give employees ownership of their jobs; and promote teamwork.

■ *Establish clear expectations.* Make sure all employees know what is expected of them as individuals and as members of the team. People need to know what is expected of them and how and to whom they are accountable for what is expected.

■ *Provide for anxiety venting.* The workplace can be stressful in even the best organizations. Deadlines, performance standards, budget pressures, and competition can all produce anxiety in employees. Consequently, leaders need to give their direct reports opportunities to vent in a nonthreatening, affirming environment. This means listening supportively. This means letting the employee know that you will not "shoot the messenger" and then listening without interrupting, thinking ahead, focusing on preconceived ideas, or tuning out.

■ *Build trust.* Negativity cannot flourish in an atmosphere of trust. Leaders can build trust between themselves and employees and among employees by applying the following strategies: always delivering what is promised; remaining open-minded to suggestions; taking an interest in the development and welfare of employees; being tactfully honest with employees at all times; lending a hand when necessary; accepting blame, but sharing credit; maintaining a steady, pleasant temperament even under stress; and making sure that criticism is constructive and delivered in an affirming way.

■ *Involve employees.* It's hard to criticize the way things are done when you are part of how they are done. Involving employees by asking their opinions, soliciting their feedback, and making them part of the solution are some of the most effective deterrents to and cures for negativity in organizations.

TERRITORIAL BEHAVIOR IN ORGANIZATIONS

Territory in the workplace tends to be more a function of psychological boundaries than of physical boundaries. In her book *Territorial Games*, Annette Simmons says, "The territorial impulse is deeply rooted in our survival programming. We are territorial because territory helps us survive. It did so thousands of years ago and it still does today. If you look at it backwards, survival needs started the whole concept of territory. The problem now may be that we are still using old territorial behaviors that are no longer appropriate to our new environment."[16]

Manifestations of Territoriality

The territorial instinct shows up in a variety of ways in an organization. Simmons lists the following manifestations:[17]

■ *Occupation.* These games include actually marking territory as "*mine*"; playing the gatekeeper game with information; and monopolizing resources, information, access, and relationships.

■ *Information manipulation.* People who play territorial games with information subscribe to the philosophy that information is power. To exercise power, they withhold information, bias information to suit their individual agendas (spin), cover up information, and actually give out false information.

■ *Intimidation.* One of the most common manifestations of territoriality is intimidation—a tactic used to frighten others away from certain turf. Intimidation can take many different forms, from subtle threats to blatant aggression (physical or verbal).

■ *Alliances.* Forming alliances with powerful individuals in an organization is a commonly practiced territorial game. The idea is to say without actually having to speak the words that "you had better keep off my turf, or I'll get my power friend to cause trouble."

- *Invisible wall.*   Putting up an invisible wall involves creating hidden barriers to ensure that a decision, although already made, cannot be implemented. There are hundreds of strategies for building an invisible wall, including stalling, losing paperwork, forgetting to place an order, and so on.
- *Strategic noncompliance.*   Agreeing to a decision up front with no intention of carrying the decision out is called strategic noncompliance. This tactic is often used to buy enough time to find a way to reverse the decision.
- *Discrediting.*   Discrediting an individual as a way to cast doubt on that person's recommendation is a common turf protection tactic. Such an approach is called an ad hominem argument, which means if you cannot discredit the recommendation, try to discredit the person making it.
- *Shunning.*   Shunning, or excluding an individual who threatens your turf, is a common territorial protection tactic. The point of shunning is to use peer pressure against the individual being shunned.
- *Camouflaging.*   This tactic is also referred to as *throwing up a smoke screen* or *creating fog.* It involves confusing the issue by raising other distracting controversies, especially those that will produce anxiety, such as encroaching on turf.
- *Filibustering.*   Filibustering means talking a recommended action to death. The tactic involves talking at length about concerns—usually inconsequential—until time to make the decision runs out.

Overcoming Territorial Behavior

Overcoming territorial behavior requires a two-pronged approach: (1) recognize the manifestations described earlier and admit that they exist, and (2) create an environment in which survival is equated with cooperation rather than territoriality. Simmons recommends the following strategies for creating a cooperative environment:[18]

- *Avoid jumping to conclusions.*   Talk to employees about territoriality versus cooperation. Ask to hear their views, and listen to what they say.
- *Attribute territorial behavior to instinct rather than people.*   Blaming people for following their natural instincts is like blaming them for eating. The better approach is to show them that their survival instinct is tied to cooperation, not turf. This is done by rewarding cooperation and applying negative reinforcement to territorial behavior.

- *Ensure that no employee feels attacked.* Remember that the survival instinct is the motivation behind territorial behavior. Attacking employees, or even letting them feel as if they are being attacked, will only serve to trigger their survival instinct. To change territorial behavior, it is necessary to put employees at ease.

- *Avoid generalizations.* When employees exhibit territorial behavior, deal with it in specifics as opposed to generalizations. It is a mistake to witness territorial behavior by one employee and respond by calling a group of employees together and talking about the issue in general terms. Deal with the individual who exhibits the behavior and focus on specifics.

- *Understand "irrational" fears.* The survival instinct is a powerful motivator. It can lead employees to cling irrationally to their fears. Managers should consider this point when dealing with employees who find it difficult to let go of survival behaviors. Be firm but patient, and never deal with an employee's fears in a denigrating or condescending manner.

- *Respect each individual's perspective.* In a way, an individual's perspective or opinion is part of his or her psychological territory. Failure to respect people's perspectives is the same as threatening their territory. When challenging territorial behavior, let employees explain their perspectives and show respect for them, even if you do not agree.

- *Consider the employee's point of view.* In addition to giving an appropriate level of respect to an employee's perspectives, managers should also try to "step into the employee's shoes." How would you as a manager feel if you were the employee? Sensitivity to the employee's point of view and patience with that point of view are critical when trying to overcome territorial behavior.

LESSONS ON CONFLICT MANAGEMENT FROM SELECTED LEADERS

Following are excerpts from the lives of several leaders that exemplify some of the conflict management principles set forth in this chapter. The leaders selected for inclusion are Abraham Lincoln, Charles Wang, Robert E. Lee, and Victoria Sandvig.

Abraham Lincoln

Few leaders have had to deal with so much conflict under more trying circumstances than Abraham Lincoln. Consider his situation. At various times during the Civil War, Lincoln had to deal with generals who were more interested in self-promotion than with winning the war; with a nation divided against itself; with cabinet members who often disagreed with him and even schemed against him; with congressmen who thought he never should have been president in the first place; and with other politicians who wanted his job. With all of the constant and continual conflict to deal with, Lincoln could be forgiven for occasionally losing his temper. However, unlike lesser leaders, Lincoln found a positive way to *blow off steam* out of sight of the other parties to the conflict. The conflict management lesson from Abraham Lincoln is as follows:

> When involved in a conflict situation, the leader who loses his temper loses *period*. A leader cannot effectively manage conflict if he cannot manage his own temper.

Lincoln's advice to people concerning controlling one's temper was simple. When angry, sit down and write the person you are angry with the harshest possible letter. Put all of your anger, venom, and frustration into the letter. Then let the letter sit overnight. In the morning, throw it away.

Lincoln often gave this advice, but he also followed it. This allowed him to deal with problems in a calm frame of mind rather than trying to do so while in a mental fog induced by agitation, frustration, and anger. "Recall that he (Lincoln) drafted a statement refuting McClellan's slanderous remarks concerning the 'Antietam episode' but would not allow Ward Lamon to deliver it. And to General George Meade after Gettysburg Lincoln wrote a scathing letter discussing his dissatisfaction with the general for not engaging Lee's army and for allowing their escape into Virginia. But Lincoln would not send this letter either."[19]

Another version of Lincoln's approach to managing his own temper so that he would be better able to manage conflict is to just bite your tongue and back off until you have had time to settle down emotionally. Anger is an emotion. Consequently, given time it will pass. Lincoln's strategy of writing down his frustrations and angry thoughts offers the added advantage of "getting it out of your system." There is one caution to be observed when following Lincoln's advice: beware of email. If you write down your angry thoughts using email, it is just too easy to click the send key. Once sent, an email message cannot be retrieved. If you follow Lincoln's advice, write your angry thoughts out in longhand or in a word-processing program.

Charles Wang[20]

As CEO of Computer Associates International, Inc., Charles Wang had his share of conflicts to deal with. Computer Associates is one of the largest independent computer software companies in the world. Wang's is a fast-paced business that operates in a highly competitive global environment. Conflicts about strategies, tactics, and decisions are inevitable in such a setting. The conflict management lesson taught by Charles Wang is as follows:

> To manage conflict effectively, a leader must be viewed by all involved as
> a person of integrity.

Wang has some definite ideas about the concept of integrity. He thinks succeeding as a CEO and exemplifying integrity go hand in hand. Wang comes from the old school in which a person's word is his bond. He thinks that the higher a leader goes in an organization, the more important the concept of integrity becomes, because the higher one goes, the greater the visibility and public scrutiny. Doing the right thing, even when people aren't looking, is not just good ethics, it's good business.[21]

One of the principal causes of conflict in the workplace is the fear that someone has a personal agenda that will interfere with "my needs." When it is known that the leader who is trying to manage or resolve a conflict can be trusted, this fear is mitigated. Being a leader in a situation in which members of an organization are in conflict is a lot like being a referee in a boxing match. Both sides have to trust that you are objective, unbiased, and trustworthy.

Robert E. Lee

When people think of Robert E. Lee, they typically think of the grandfatherly Lee who so brilliantly led the Army of Northern Virginia through most of the Civil War, outthinking, outmaneuvering, and outfighting better-equipped, better-provisioned, but not better-led Union forces. However, before Lee became the dashing figure most people remember, he spent many years toiling away as a military engineer. An incident that occurred during this time when Lee was serving as an engineer teaches an important conflict management lesson:

> Leaders who fail to effectively manage conflict are at risk of losing their
> best and brightest personnel.

In the case in question, it was Lee—a man who became one of the most effective, most celebrated military tacticians ever to lead an army—who was almost lost because leaders above him failed to effectively deal with conflict. "The Corps of Engineers had considerable autonomy in contracting for labor and materials. Since other army units had no such latitude, this became a source of resentment and friction. During Lee's time at Fort Monroe, there were several disputes and almost constant sniping between the engineers and the rest of the garrison. This so discouraged Lee that he requested a transfer."[22]

In fact, Lee not only requested a transfer because of the constant squabbling at Fort Monroe but also gave serious consideration to leaving the army altogether. Concerning conflict on the job, Al Kaltman has this to say: "Inter-unit disputes demoralize people. Constant turf squabbles will cost you your best people. As a manager, it's your responsibility to set a cooperative tone. Avoid making remarks and taking actions that foster petty jealousies and unprofessional attitudes and conduct. If you see a quarrel developing between your people and those from another unit, don't let it fester. Nip it in the bud."[23]

Victoria Sandvig[24]

One of the most effective conflict management strategies is to establish a common vision among all employees—let them see the big picture and where they fit into it. Of course, this is easier said than done. Even in small companies, it can be difficult to get all employees "singing from the same page." Just imagine how difficult it would be to establish a common vision among thirteen thousand employees. This was the challenge faced by Victoria Sandvig, vice president of the Event and Production Services Department of Corporate Communications at Charles Schwab & Company, Inc. The conflict management lesson taught by the career of Victoria Sandvig is as follows:

> To prevent and manage conflict, leaders must establish a common vision among stakeholders. To do this may require a creative, innovative approach.

Sandvig accomplished the daunting challenge of establishing a common vision among all of Charles Schwab's employees with a program that came to be known as VisionQuest.

> The beginnings of VisionQuest stirred when a group called Root Learning, Inc., started creating some maps of vision and values, originally for the executive committee. As Victoria puts it, "One of these maps addressed questions of who the company is, its vision, and values. The second one was very much about where we want to go, who our competition is, who our clients are, the profile of our customer." After the executive committee, the maps were presented to the senior management team, where their power became obvious to the larger group. Then began the process of involving the company at large. ... And so, across the United States and across the world, Schwab employees gathered on a Saturday. ... The six-hour event kicked off in San Francisco's Moscone Center at 8:30 A.M., with six thousand people there. Other locations had fewer people—and they were all hooked up via satellite.[25]

Whether at the Moscone Center in San Francisco or at other locations around the world, Schwab's employees sat in groups of 10; mixed according to tenure, position, titles, and so forth. For six hours, Schwab's employees at all levels discussed their company's vision, values, direction, and so on. As a result, in just one day, all 13,000 Schwab employees established a common understanding of the company's vision and corporate beliefs. Now when differences of opinion arise concerning solutions to problems or what strategies to adopt, leaders have a common vision to use in resolving the conflict.

Summary

1. Internal politics consists of activities undertaken to gain advantage or influence organizational decision making in ways intended to serve a purpose other than the best interests of the overall organization. It is the games people play to promote decisions that are based on criteria other than merit.

2. The most commonly used methods of internal politicians are lobbying, building coalitions, applying harassment and pressure, electioneering, gossiping, and spreading rumors.

3. The rationale for collaboration is found in the debilitating effect internal politics can have on an organization. Internal politics can drain an organization of its intellectual and physical energy and in the process take away its ability to compete.

4. An organization's effort to control internal politics should have at least the following components: strategic planning, leadership, reward and recognition, performance appraisal, customer focus, conflict management, and cultural.

5. The most common categories of negative behavior are control disputes, territorial (boundary) disputes, dependence and independence issues, need for attention and responsibility, authority, and loyalty issues.

6. The following symptoms are indicators of negativity in the workplace: "I can't" attitudes, "they" mentality, critical conversation, and blame fixing among employees.

7. To overcome negativity, organizations should communicate, establish clear expectations, provide opportunities for anxiety venting, build trust, and involve employees.

8. Territorial behavior in organizations manifests itself in the following ways: occupation, information manipulation, intimidation, alliances, invisible walls, strategic noncompliance, discrediting, shunning, camouflaging, and filibustering.

9. The following strategies will help when trying to overcome territorial behavior: avoid jumping to conclusions, attribute the behavior to instinct rather than people, ensure that employees don't feel attacked, avoid generalizations, understand irrational fears, respect each individual's perspective, and consider the employee's point of view.

10. Causes of workplace conflict include limited resources, incompatible goals, role ambiguity, different perceptions, and poor communication.

11. Managers have two responsibilities regarding conflict in the workplace: conflict resolution and conflict stimulation. Conflict should be stimulated to overcome excessive compliance and complacency.

Key Terms and Concepts

Alliances	Information manipulation
Ambition	Internal politics
Anxiety venting	Intimidation
Blame fixing	Invisible walls
Building coalitions	Leadership
Camouflage	Lobbying
Conflict management	Need for acceptance
Critical conversation	Occupation
Culture	Performance appraisal
Customer focus	Personal insecurity
Ego	Reward and recognition
Electioneering	Self-interest
Gossiping and spreading rumors	Shunning
Harassment and pressure	Strategic noncompliance
Hunger for power	Strategic planning
"I can't" attitudes	"They" mentality

Review Questions

1. Define *internal politics.*
2. List and briefly describe the most commonly used methods of internal politicians.
3. Describe the impact internal politics can have on an organization's quality and competitiveness.
4. Describe how managers can control internal politics in an organization.
5. What are the categories of negativity in the workplace?
6. Explain the strategies for overcoming territorial behavior.
7. When should conflict be encouraged in an organization?

LEADERSHIP SIMULATION CASES

The following simulations are provided to generate additional thought and discussion about the principles of leadership explained in this chapter. Readers are encouraged to consider how the situations presented in these cases might apply to them and to discuss the cases with other leaders and leadership candidates.

CASE 10.1 I Don't Understand Why We Have So Much Conflict

David Ord had been in his first leadership position for less than a month when he became concerned about what he considered an inordinate amount of conflict in his department. When he had been a design engineer with Davidson Engineering Services (DES), Ord was aware of the occasional conflict. In fact, he had even been involved in a few himself. But now that he was the director of the design department at DES, Ord felt almost overwhelmed by the constant conflicts among his best and brightest personnel. Out of frustration, he decided to talk with a colleague who had been in upper management at DES for more than 10 years.

"Mark, I don't understand why we have so much conflict in my department. I have good people who are good at their jobs, but I can't even get to the paperwork on my desk because I am constantly having to referee disputes," claimed a frustrated David Ord. His more experienced colleague, Mark Vetak, gave Ord a look that said, "Go ahead, I'm listening?" When Ord had finished venting, he ask Vetak, "Why do you think good people like mine are so quick to disagree and be disagreeable?"

Discussion Questions

1. Have you ever worked in a setting in which there seemed to be a lot of conflict? Is so, how did it affect your morale?
2. Put yourself in Mark Vetak's place. How would you answer Ord's question? What are some of the factors that could be causing conflict in his department?

CASE 10.2 Let 'Em Fight. What Can It Hurt?

John Evan had been a drill instructor in the United States Marine Corps when he was a young man. Although that had been many years ago, he still thought like a drill instructor. Consequently, when Susan Beaton, one of his colleagues at Bryon Manufacturing Services, asked for Evan's advice about how to deal with conflict in her division, Evan surprised her by saying, "Let 'em fight. What can it hurt?" "You must be kidding," said an astonished Beaton. "It's hurting our productivity plenty, John."

"That's only temporary," said Evan reassuringly. "Leave them alone and they will eventually work it out. They're adults." When Beaton still wasn't convinced, Evan said, "Let me tell you a story from my days as a drill instructor. On Parris Island, we always had a few hotheads who were quick to get into fights. Some drill instructors were quick to rush in and break up these fights, but not me. I knew better. I simply told everybody to stand back and let them go at it. One of the reasons these hotheads were so quick to fight is they thought someone would break it up before anyone got hurt. Let me tell you, Susan—when they found out that in my platoon they would have to fight it out until one or both couldn't fight anymore, I never had another problem with fighting. On the other hand, those drill instructors who were quick to break up fights had to deal with fighting every day."

Discussion Questions

1. What do you think of Evan's approach to handling conflict? Apparently it worked in Marine Corps boot camp. Do you think it would work on the job?
2. Do you agree with Evan's comment, "Leave them alone and they will eventually work it out. They're adults."

CASE 10.3 This Work Environment Is So Negative It's Toxic

Andrea Cole had never worked in a situation that was so persistently negative. When an old college friend called to congratulate her for beating out a strong field to be the new vice president for the technology department at Global Technologies, Inc., Cole responded by saying, "I'm not sure congratulations are in order. I have never worked in a place where everyone is so negative. There is constant bickering, backbiting, and finger-pointing. This work environment is so negative it's toxic."

Trying to encourage her, Cole's friend said, "Well, Andrea, just make overcoming all of that negativity your first order of business. You can do it. I know you can. I've seen you in action." "I hope so," said Cole. "Negativity is like a disease in this place."

Discussion Questions

1. Have you ever worked in an environment that was consistently negative? If so, how did the negativity affect your work and that of others?

2. How would you advise Andrea Cole to go about overcoming the negativity in her organization?

Endnotes

[1] Louis E. Boone, *Quotable Business*, 2nd ed. (New York: Random House, 1999), 57.

[2] K. W. Thomas, "Conflict and Conflict Management," *Academy of Management Review* 2 (1997): 163.

[3] K. W. Thomas, "Toward Multi-Dimensional Values in Technology," *Academy of Management Review* 2 (1977): 487.

[4] S. P. Robbins, "Conflict Management and Conflict Resolution Are Not Synonymous Terms," *California Management Review* 32 (1978): 67–75.

[5] D. Tjosvold, "Making Conflict Productive," *Personnel Administrator* 29 (1984): 121–130.

[6] S. P. Robbins, 67–75.

[7] S. P. Robbins, 67–75.

[8] James R. Wilcox, Ethel M. Wilcox, and Karen M. Cowart, "Communicating Creatively in Conflict Situations," *Management Solutions: Special Reports* (New York: AMACOM, 1987), 34–40.

[9] Douglas MacArthur, *Reminiscences* (New York: McGraw Hill, 1964), 184.

[10] Alan R. Gitelson, Robert L. Dudley, and Melvin J. Dubnick, *American Government* (Boston: Houghton Mifflin, 1993), 225.

[11] Jay A. Conger and Beth Benjamin, *Building Leaders* (San Francisco: Jossey-Bass, 1999), 100.

[12] Jay A. Conger and Beth Benjamin, 100.

[13] Bob Nelson, *101 Ways To Reward Employees* (New York: Workman, 1994), 28.

[14] Tom Rusk, *The Power of Ethical Persuasion* (New York: Penguin, 1994), xvi–xvii.

[15] *The American Heritage College Dictionary*, 3rd ed, S.V. "culture."

[16] Annette Simmons, *Territorial Games* (New York: AMACOM, 1998), 6–7.

[17] Annette Simmons, 179.

[18] Annette Simmons, 187.

[19] Donald T. Phillips, *Lincoln on Leadership* (New York: Warner Books, 1993), 80–81.

[20] Thomas J. Neff and James M. Citrin, *Lessons from the Top* (New York: Currency/Doubleday, 2001), 327–332.

[21] Thomas J. Neff and James M. Citrin, 330.

[22] Al Kaltman, *The Genius of Robert E. Lee* (Upper Saddle River, NJ: Prentice Hall, 2000), 23.

[23] Al Kaltman, 23.

[24] James M. Kouzes and Barry Z. Posner, *The Leadership Challenge*, 3rd ed. (San Francisco: Jossey-Bass, 2002), 162–163.

[25] James M. Kouzes and Barry Z. Posner, 163.

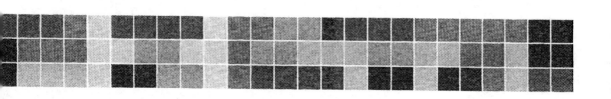

Index